This important book brings recent findings and theories in biology and psychology to bear on the fundamental question in ethics of what it means to behave morally. It explains how we acquire and put to work our capacities to act morally and how these capacities are reliable means to achieving true moral beliefs, proper moral motivations, and successful moral actions.

By presenting a complete model of moral agency based on contemporary evolutionary theory, developmental biology and psychology, and social cognitive theory, the book offers a unique perspective. It will be read with profit by a broad swathe of philosophers, as well as psychologists and biologists.

T0245164

The biology and psychology of moral agency

CAMBRIDGE STUDIES IN PHILOSOPHY AND BIOLOGY

General Editor
Michael Ruse, *University of Guelph*

Advisory Board
Michael Donoghue, *Harvard University*
Jean Gayon, *University of Bourgogne*
Jonathan Hodge, *University of Leeds*
Jane Maienschein, *Arizona State University*
Jesus Mosterin, *University of Barcelona*
Elliott Sober, *University of Wisconsin*

This major series publishes the very best work in the philosophy of biology. Nonsectarian in character, the series extends across the broadest range of topics: evolutionary theory, population genetics, molecular biology, ecology, human biology, systematics, and more. A special welcome is given to contributions treating significant advances in biological theory and practice, such as those emerging from the Human Genome Project. At the same time, due emphasis is given to the historical context of the subject, and there is an important place for projects that support philosophical claims with historical case studies.

Books in the series are genuinely interdisciplinary, aimed at a broad cross section of philosophers and biologists, as well as interested scholars in related disciplines. They include specialist monographs, collaborative volumes, and – in a few instances – selected papers by a single author.

Published Titles
Alfred I. Tauber, *The Immune Self: Theory or Metaphor?*
Elliott Sober, *From a Biological Point of View*
Robert Brandon, *Concepts and Methods in Evolutionary Biology*
Peter Godfrey-Smith, *Complexity and the Function of Mind in Nature*

The biology and psychology of moral agency

WILLIAM A. ROTTSCHAEFER

Lewis and Clark College

CAMBRIDGE
UNIVERSITY PRESS

CAMBRIDGE UNIVERSITY PRESS
Cambridge, New York, Melbourne, Madrid, Cape Town, Singapore, São Paulo

Cambridge University Press
The Edinburgh Building, Cambridge CB2 8RU, UK

Published in the United States of America by Cambridge University Press, New York

www.cambridge.org
Information on this title: www.cambridge.org/9780521592659

First published 1998
This digitally printed version 2008

A catalogue record for this publication is available from the British Library

Library of Congress Cataloguing in Publication data
Rottschaefer, William A. (William Andrew), 1933–
The biology and psychology of moral agency / William A.
Rottschaefer.
p.   cm. – (Cambridge studies in philosophy and biology)
Includes bibliographical references and index.
ISBN 0-521-59265-8
1. Ethics, Evolutionary.   I. Title.   II. Series.
BJ1311.R68   1997
171′.7–dc21   97-8764
                                                          CIP

ISBN 978-0-521-59265-9 hardback
ISBN 978-0-521-06450-7 paperback

Contents

v

Contents

Preface

In this book I propose a scientific naturalistic account of moral agency, offering answers to four central questions: (1) what counts as moral agency, both substantively and functionally? (2) how do we acquire our capacities as moral agents? (3) how do we put these capacities to work? and (4) what makes for justified true moral beliefs, proper moral motivations, and successful moral action? I argue that moral agency is a phenomenon of the natural world best understood with the help of sciences. Making use of recent theories and findings in evolutionary theory, developmental biology and psychology, and social cognitive theory in psychology, I set forth a model of moral agency as a complex four-level capacity consisting of (1) a base level of both evolutionarily based and operantly learned capacities; (2) a behavioral level consisting of cognitively acquired moral beliefs and desires that is the immediate source for moral behaviors; (3) a reflective level composed of moral beliefs and desires concerning the behavioral-level moral beliefs and desires and regulative of the latter; and (4) a self-referentially reflective level by means of which an agent conceives of herself as a moral agent.

In proposing my model, I pursue a goal common to many philosophers, the search for what Wilfrid Sellars (1963) aptly called the *synoptic vision:* the attempt to see things as a whole. Sellars's goal was to articulate the connections between what he termed the *manifest* and *scientific images* of human persons in order to achieve a synoptic vision. It was my attempt to work through Sellars's monumental efforts to do this that brought my previous philosophical endeavors to a focus and gave them subsequent direction. Although trained in philosophy of science and physics, the discipline to which philosophers of science have devoted so much of their analytic and synthetic efforts, I found myself teaching at a small liberal arts college with colleagues in psychology and biology who had great interest in both the substantive philosophical issues connected with their disciplines and the

ix

contributions that philosophy of science could make to an understanding of their disciplines. The issues we discussed often focused on questions concerning science and values. Thus began my journey through the highways and byways of behaviorism, B. F. Skinner's science of values, and cognitive social learning theories. The intrinsic connections between psychology and biology, as well as the emergence of sociobiology into popular and philosophical consciousness, next led to a study of E. O. Wilson's proposal to biologize ethics and subsequent developments in evolutionary ethics. From there the path into developmental biology and psychology become a necessary one to follow.

Another strain in my philosophical education provided a guiding principle for these investigations. My mentor at Boston University, Abner Shimony, not only introduced me to the work of Sellars but demonstrated the importance of bringing to bear the theories and findings of the natural sciences, especially biology and psychology, on both the traditional problems of epistemology and the epistemological problems raised in philosophy of science. It became clear to me that a synoptic vision could not be achieved without a similar endeavor in ethics or moral philosophy. Although not a specialist in moral philosophy, I found myself involved in trying to address questions about the nature of morality and the justification of moral beliefs, motivations, and actions from a scientific naturalistic perspective. Although it is necessarily incomplete because it does not bring in the relevant contributions of the social sciences such as anthropology, sociology, political science, and economics, I offer an account of the biology and psychology of moral agency. The plausibility of this account is to be measured not only by the extent to which it is faithful to the findings of the sciences that it employs but also by the extent to which it provides a scientific naturalistic alternative to autonomous commonsense, philosophical, or religious accounts of moral agency, one that better answers the questions about moral agency.

I am very indebted to Michael Ruse, who offered much encouragement, support, and advice. I owe much to Mike Bradie and David Shaner, who read and commented on a much longer version of the manuscript. I also thank Ted Schoen, who provided helpful suggestions on another earlier version. I thank Wendell Stephenson for conversations on several sections of an earlier version. My colleagues in philosophy, Sevin Koont and Clayton Morgareidge, offered many helpful comments on earlier portions of the book that I presented at our colloquia. I owe much to my colleague in biology, Dave Martinsen, and in mathematics, Bob Owens, and to my former colleague in psychology, Bill Knowlton. I would also like to thank two anonymous referees for Cambridge University Press for their very helpful comments and suggestions.

I am grateful to my student research assistants Jesa Witherbee, Amber Ontiveros, Leah Berman, and Justin Phillips and to the editors at Cambridge University Press. I also wish to thank the publishers of *Zygon* for permission to use portions of my "B. F. Skinner and the Grand Inquisitor" (1995, pp. 407–34) in Chapters 5 and 8 and the publishers of *Behavior and Philosophy* for permission to use portions of my "Social Learning Theories of Moral Agency" (1991, pp. 61–76) in Chapter 6.

Finally, my special appreciation goes to my wife, Marie Schickel Rott-schaefer, without whom this book would not be.

Introduction

In this study I propose an account of the biological and psychological bases of moral agency. I am motivated to do so by a commitment common to many philosophers: the search for what Wilfrid Sellars aptly called the *synoptic vision,* the attempt to see things as a whole: "The aim of philosophy, abstractly formulated, is to understand how things in the broadest possible sense of the term hang together in the broadest possible sense of the term" (Sellars 1963, p. 1). In order to achieve a synoptic vision of the whole, Sellars aimed to articulate the connections between what he termed the *manifest* and *scientific* images of human persons. A guiding principle for my investigation is the new scientific naturalistic turn in philosophy, the attempt to bring to bear the best theories and findings of the sciences in the solution of philosophical problems.

The use of the sciences has immeasurably enhanced philosophical attempts to understand such phenomena as time, space, matter, motion, change, causality, and life. Today, philosophy of mind and epistemology are both feeling the positive effects of inputs from biology and the cognitive sciences. Scientifically informed philosophical investigations have, I contend, advanced the quest for a synoptic vision of things. However, I do not believe that a synoptic vision of human beings can be achieved without a similar endeavor in ethics or moral philosophy. I believe that the natural and social sciences can contribute significantly to answering traditional ethical questions about the nature and function of morality and the justification of moral claims. My goal is to set forth a proposal for a naturalistically based account of moral agency, one that makes substantive use of recent findings and theories in evolutionary theory, developmental biology, and psychology, as well as in behavioral and cognitive behavioral psychology. I make no claims to have surveyed, let alone assessed and integrated, the prodigious amount of biological or psychological literature that prima facie appears relevant to providing an account of the biological and psychological

1

bases of moral agency. Nor do I pretend to examine and assess the multitude of philosophical accounts of moral agency. The issues, approaches, and materials I take up are to some extent due to autobiographical vagaries; despite that, my hope is that they are sufficiently relevant to the development of a naturalistic account of moral agency that they will foster further thought, reflection, and development of such an account. A more complete scientific naturalistic account will require an examination of the relevant contributions of the social sciences such as anthropology, sociology, political science, and economics.

The plausibility of such scientific naturalistic accounts is to be measured not only by the extent to which they themselves offer satisfactory answers to the central questions of moral agency, but also by the extent to which they provide more satisfactory alternatives to autonomous commonsense, philosophical, or religious accounts of moral agency. In addition, a successful scientific, naturalistic account ought to provide a more satisfactory articulation of the connections between our ordinary conceptions of moral agency (roughly Sellars's manifest image) and scientific accounts than those provided by its nonnaturalistic competitors. However, for the most part, I leave such an assessment of comparative merit for another day.

The book is divided into five parts. In Part I, I outline, and set the stage for the development of, my scientific naturalistic account of moral agency. I formulate four central questions that an adequate theory of moral agency needs to answer: (1) what counts as moral agency, both substantively and functionally? (the question of relevance); (2) how do we acquire our capacities as moral agents? (the question of acquisition); (3) how do we put these capacities to work? (the question of action); and (4) what makes for justified true moral beliefs, proper moral motivations, and successful moral action? (the question of adequacy). In order to answer these questions, I propose a biologically and psychologically based model of moral agency that has four levels: (1) a base level, consisting of both biologically and psychologically based (though minimally cognitive) moral capacities; (2) a behavioral level, consisting of cognitively acquired moral beliefs and desires, that is the immediate source of moral behaviors; (3) a reflective level, composed of moral beliefs and desires concerning the behavioral-level moral beliefs and desires and regulative of the latter, as, for instance, moral norms seem to be; and, finally, (4) a self-referentially reflective level by means of which an agent conceives of herself as a moral agent. My contention is that, by using this model of moral agency, we can make substantial progress in answering the central questions about moral agency.

The proposed model instantiates an integrationist position on the question of how the sciences might contribute to answering questions about moral agency. The integrationist position I adopt holds that disciplines in biology and psychology can and ought to make significant contributions to answering questions about moral agency by (1) providing the information necessary for making moral decisions both at the applied level and at the normative level; (2) describing and explaining human moral capacities, both their acquisition and maintenance, as well as their operation; (3) providing some bases for moral norms; (4) furnishing accounts of the justification of moral beliefs, motivations, and actions, as well as a theory about moral realities; (5) enabling a critique of alternative accounts of ethics, for instance, commonsense or folk psychologically based ethics, a priori moral philosophy, or religious ethics, as well as those of reductionist or eliminativist stripes; and, finally, (6) addressing questions of the meaning of ethics from a scientific naturalistic perspective. The success of the proposed model and the provision of a satisfying synoptic vision of moral agency would lend further weight to these integrationist hypotheses about the connections between the sciences, especially biology, psychology, and ethics, and to the scientific naturalistic approach to philosophy generally.[1]

In Part II, I explore the evolutionary bases of moral agency. In Chapter 2, I distinguish three themes from evolutionary theory that link it with issues of moral agency: first, our evolutionarily fashioned human nature; second, our evolutionarily based social nature; and third, our evolutionarily based altruistic tendencies. I maintain that the findings from evolutionary theory and sociobiology are relevant to moral agency and argue for the theoretical plausibility of the existence of evolutionarily based moral capacities. Next, in Chapter 3, I proceed to the question of the actual existence and nature of such capacities, examining the contribution of E. O. Wilson. Although some biologists, for example, George C. Williams, find biology and morality to be in mortal conflict and others, like Richard Alexander, see morality to be in the service of biology, I argue that Wilson's vision of biology in the service of morality is preferable but that his proposal for a biologically based account of moral agency is flawed. Besides requiring further empirical support, it must be revised from an evolutionary perspective and supplemented by developmental and psychological perspectives. Chapter 4 is devoted to ontogeny and development. There I turn to the work of developmental psychologist Martin Hoffman on our capacity for empathic distress to outline the case for the existence of one component of a evolutionarily based but plastic moral sense, that is, one that is open to environmental input and learning.

The introduction of learning, even in the case of evolutionarily based moral capacities, leads us naturally in Part III to a consideration of the psychological bases of moral agency. I explore and assess the contribution of B. F. Skinner's "science of values" to an account of moral agency in Chapter 5 and argue that, despite its behaviorist limitations, the operantly conditioned capacities that he postulates can be incorporated at the base level of an integrationist model of moral agency. I turn in Chapter 6 to the results of the cognitive revolution in psychology and contend that their incorporation into an integrationist account of moral agency requires the introduction of a cognitive-behavioral level into our model of moral agency. In doing so, I must meet the challenge of what I call an *investigative dilemma:* The most adequate theories of acquisition and action, the so-called social learning theories, seem to be morally irrelevant, that is, they do not appear to be about morality at all, and the psychological theories that seem to be most clearly about moral agency, specifically, the cognitive developmental theories, appear to be unable to answer questions concerning the acquisition and enactment of moral agency. So an integrationist is faced with the charge of either proposing theories of agency that are irrelevant to moral agency or proposing a theory of moral agency that seems to be genuinely about morality but fails to answer in an adequately scientific fashion the questions of acquisition and action. I resolve the investigative dilemma by taking up the side of cognitive social learning theories, in particular Albert Bandura's social cognitive theory, and argue that the charge of irrelevance fails. In so doing, I complete the presentation of my scientific naturalistic model of moral agency, showing that Bandura's theory provides scientific support for the reflective and self-referentially reflective levels of moral agency.

In Chapter 7, I address a second challenge to my integrationist approach, what I call the *reductionist predicament.* It maintains, on the one hand, that any adequately scientific account of human agency may in fact demand the elimination of factors apparently necessary for moral agency, at least moral agency conceived in folk-psychological terms, that is, in terms of beliefs and desires. On the other hand, if the integrationist fails to make use of adequately supported scientific categories and theories, she is left with the uninviting prospect of abandoning her scientific and naturalistic commitments in order to hold on to an adequate model of moral agency. I consider some findings in the neurosciences about the neurophysiological bases of our cognitive and agential capacities and the implications of these findings for the reduction or elimination of our folk-psychological categories of agency. In response, I elaborate an emergentist account of moral agency and argue that currently, at least, such an account can hold its own empirically against reductionist and eliminativist counterproposals. Thus, I contend that the integrationist emerges from the trials of the reductionist predicament with a modest confidence that she has an account of moral agency that answers the questions of action and acquisition with relative scientific adequacy, given the current state of development in the relevant scientific disciplines. Moreover, I hold that she has met the challenge of irrelevance by clearly exhibiting that her scientifically elaborated account of agency is indeed about moral agency. In terms of the substantive connections between the sciences and ethics, I conclude that I have gone a long way toward confirming an explanatory connection, as well as illustrating the way that an integrationist account can enable critiques of alternative accounts of moral agency.

Part IV addresses the last of the major questions about moral agency: the question of adequacy. Answering this question enables the integrationist, in turn, to support the metaethical connection hypothesis. Drawing on recent work in naturalized epistemology, I argue in Chapter 8 that justifications of moral beliefs, motivations, and actions should be conceived of in terms of reliable mechanisms. I illustrate this approach by making use of some of the mechanisms of acquisition and action already laid out in answering the questions of acquisition and action. Answering the question of adequacy also supports the epistemological component of my case for the metaethical connection hypothesis and leads naturally to the ontological side of that hypothesis and the question of the nature of moral realities. I address that question in Chapter 9, taking up the controversial path of moral realism and contending that moral realities are complex relational realities composed of moral agents, objective moral values, and the causal relationships that unite them. Discussion of ontological issues completes my case for the explanatory and metaethical connection hypotheses.

I take up the critical connection and meaningfulness hypotheses in Chapter 10, which constitutes Part V. With regard to the former, I contend that although separatist worries have not been entirely silenced, the main threat to the integrationist position comes from eliminativism. So I once again address these concerns, making use of Wilfrid Sellars's idea of the synoptic vision and the way that representations of persons as intentional agents differ from descriptions and explanations of human cognitive capacities. I argue that even though folk-psychological accounts of the latter may be eliminated, that sort of elimination does not touch human agential capacities qua agential. I address the meaningfulness hypothesis by proposing a scientific naturalistic theory of meaningfulness and comparing it with separatist accounts of the meaningfulness of moral agency stemming from both religious and humanistic perspectives.

While recognizing the preliminary and incomplete character of these efforts, I conclude that a scientific naturalistic model of moral agency of the sort that I have proposed is a plausible one and worthy of further investigation, and that the integrationist hypotheses on which the model is built provide a fruitful approach to the investigation of moral agency. By these means, I hope to have extended the synoptic vision concerning human persons and to have given further support to a scientific naturalistic pursuit of this vision.

NOTE

1. I shall assume that the informational connection is unproblematic and, although the question of whether findings from biology and psychology can provide substantive input to moral norms is an important issue for any integrationist account, I shall not address it here. To that extent, my account remains significantly incomplete.

I

Moral agency and scientific naturalism

1

Understanding moral agency

What is a scientific naturalistic view of moral agency?

1.1 MORAL AGENCY

We begin by considering an ordinary perspective on being a moral agent. By an ordinary perspective I mean one that is shaped by our ordinary experience of and reflection on being a moral agent. You can probably recall many small and some very large moral decisions in your life. Some of these are more personal – regarding, for instance, telling the truth in a situation where the truth was rightfully demanded, but it was in your interest to be silent or to shade your story. Some decisions may have been more communal – whether, for instance, your professional group or union should actively oppose or support the right of a woman to have an abortion. Many decisions, no doubt, concerned issues in which science and technology played some part in both the problem and possible solutions. Should I carpool? Should I support a movement in Congress to eliminate funding for environmental protection? What does the ordinary perspective on moral agency say about moral agency given experiences of the preceding sort and reflection on them? It probably says a lot of different things since, on one level, there are as many experiences and reflections as experiencers and persons reflecting on their experiences. However, let's generalize and see if we can say some broad things about moral agency.

On one ordinary view of moral agency, a person acts in a morally correct fashion when she acts on the basis of adequate moral beliefs correctly applied to a particular situation. So, for instance, you have decided that even though it costs you some valuable time, the right thing to do is to form a carpool to get to work and back. You base your decision on the importance of improving the air quality in your city. You believe that you have an obligation to do something in that regard since air quality is important for the public and for the living things in the city. Your time is also important, but not so important that the loss of the 15 to 30 minutes a day that the

9

carpooling would involve morally outweighs the contribution to air quality brought about by the use of three fewer autos a day for five days a week.

The use of moral beliefs to guide action serves as a mark of moral agency. One can raise a number of questions about moral agency so described: (1) How does one acquire moral beliefs? (2) How does one put these moral beliefs to work? (3) How does one assess the adequacy of these beliefs and actions? (4) What makes action on the basis of moral belief the relevant criterion for the identification of moral agency? These questions concern the issues of acquisition, action, justification, and relevance, respectively.

Going back to our example, we can ask ourselves these same questions about that particular moral decision. How did we come to the moral belief that the moral goods of carpooling outweighed the moral goods of individualized transportation? How did we acquire the belief that preservation of air quality was morally valuable or, more fundamentally, that it was morally wrong to harm the health of others? We can also inquire about our motivations. Let's say that motivations are roughly what can lead us to act on our moral beliefs. There are many things that we believe but that do not motivate us to do anything. Someone may know that air pollution harms the health of many people. He may also realize that he is adding to air pollution by driving to work and back every day all by himself. He may even be aware that there are alternatives, carpools, that he could join. But none of this moves him to action. So we have to ask ourselves not only how we come to have certain moral beliefs, but what it is that moves us to act on our beliefs. Is it something about a moral belief that makes it motivating in itself or do we need something else besides our moral convictions to move us to act on them?

But even supposing that we have acquired a set of moral beliefs and have been motivated to act on them, we can still raise the question of what justifies our action. Is what we are doing by carpooling the right thing to do in the circumstances? If so, how do we know that it is? Suppose we are asked to justify our decision. A critic may challenge some of the facts about our decision, the relative harmfulness of auto emissions, for instance. Do we have our facts correct? Or she may challenge our ethical reasoning, arguing that we have no obligation, at least at present, to make such a major sacrifice of time. More radically, she may argue that we have no definite obligations to help others preserve their health, and that even if others are polluting the air, we certainly don't have to stop doing so until they stop too. These questions and challenges raise questions about the correctness of our moral decision. They belong to the issue of justification. So, just as moral agency requires an account of how we developed our moral beliefs and what enables us to put them into action, it also requires that we attempt to show that these beliefs, and the practices that flow from them, are, to the best of our knowledge, justified, that is, morally right.

But all of these questions presuppose that we know what counts as being in the moral realm. Someone might agree with our decision about carpooling but think that we shouldn't really speak of it as a moral decision. Morality has to do, she might argue, with interpersonal relationships, not with public policy issues like pollution and carpooling. Those issues may belong to the legal realm. Perhaps there could be a law that carpooling is required and one could break the law by failing to do so. But such an infringement would be illegal, not immoral. Similarly, someone could argue that preserving the environment is a matter of prudence and aesthetics. We should not waste resources, especially if they are becoming scarce. To do so would be imprudent. We would not be seeing to our future in a way that is more beneficial to us and our children. In addition, the destruction of the environment that air pollution entails detracts from its aesthetic value. So if carpooling, in fact, helps to preserve the environment, then it is the aesthetic thing to do but not the moral thing to do. These reflections lead us to conclude that in order to understand moral agency, we need to have some grasp of what belongs to the moral realm and what doesn't. It will not do merely to distinguish the morally right and wrong thing to do, something that we think we have accomplished, provided that we have some justification for what we have decided to do. Important as this is, there is the prior question of what belongs in the moral realm, whether something is amoral or nonmoral, as opposed to whether it is a morally correct or incorrect action. Some legal matters may touch on moral issues, but others may be purely legal – for instance, that you need a notary public to certify that you have properly signed a document of a sale of property.

1.2 THE SCIENCES AND THE THEORETICAL PROBLEMS OF MORAL AGENCY

These questions about moral agency, about acquisition, action, justification, and relevance, certainly are not new. We have as a species been engaged in trying to clarify them and suggest answers to them for a long time. Both ordinary moral agents and philosophers have been involved in this task. Ordinary experience and reflection suggest a variety of answers to these questions. Philosophical theories attempt to clarify and refine those answers, as well as determine which ones might be preferable. For instance, we need to learn to be moral agents, yet the precise roles of nature and nurture in the acquisition of moral beliefs and their implementation are not discernible on the basis of ordinary observation and reflection. The question of acquisition remains open, and space abounds for philosophical analysis and speculation. Ordinary observations and reflections also provide us with a variety of initial answers to the issue of action. Sometimes we feel that adequate knowledge is sufficient to lead to moral action, and we lean toward the Platonic position that knowledge is virtue. On the other hand, sometimes we believe we know what we ought to do and yet fail to do it. So we tend to the Aristotelian position that knowledge of moral principles is not sufficient for moral action and thus for virtue. But neither common sense nor philosophical analysis provides a definitive solution. In like manner, with regard to relevance, we think that one who does not know what she is doing cannot be held morally responsible. Yet we have our doubts about the virtue of a father, for instance, who must explicitly reason out from moral principles that he ought to care for his ailing child. We seem to allow cognitive involvement ranging from feeling to reason as sufficient to identify an action as moral. Again, philosophical speculation is plentiful, refining and attempting to support various commonsense conceptions. Finally, common sense offers a range of sources for justifying our moral principles: for instance, feeling, authority, human nature, God, or reasoning supported by abstract principles. Philosophers have used these suggestions and others to build theories to answer the question of justification.

Thus, commonsense observations and reflection suggest a variety of answers to questions about acquisition, action, relevance, and justification. These are refined by careful philosophical analysis and explored in disciplined philosophical speculation. But in each case, it seems to me, the range of plausible solutions can be narrowed and their adequacy better assessed if we call on our best scientific knowledge for assistance. My hypothesis is that the natural and social sciences have much of importance to contribute to our understanding of ourselves as moral agents. More specifically, I am concerned with the contribution that the sciences of biology and psychology might make to our understanding of ourselves as moral agents.

It is reasonable to think that the natural and social sciences might be able to contribute to the resolution of some of the problems of moral agency, since they represent the best empirical knowledge that we have about ourselves and the world. So, even more, we might think that biology and psychology would be of particular importance in this pursuit. But even if this seems to be a plausible supposition to make, our initial reflections may make us pause and reassess things. Since, in our practical moral lives, science and technology present themselves as Janus-faced, both as a threat and as a promise, we need to raise the question about whether we find the same phenomenon on the theoretical level. Indeed, we have good reason to be cautious.

Consider three serious problems with attempting to link the sciences to ethics in order to help us understand and explain moral agency and discern the significance of our existence as moral agents. The first is the problem of freedom versus determinism; the second, facts versus values; and the third, information versus meaningfulness.

When we think about science, we often think of laws, theories, predictions, and explanations. Without, at this point, going into an analysis of these concepts, one common note that we often associate with these concepts is that of determinism. We think scientists are able, on the bases of laws and theories, to predict and explain events. And we believe they can do so because they have been able to find the causal factors that bring about an event. On the other hand, from our ordinary perspective on human actions, and moral agency in particular, we think that one of its distinctive notes is freedom. Humans are able, we believe, to make decisions for themselves. They are not always forced by the external natural or social environment or compelled by their internal physiological makeup or psychological state to do the things they do. Indeed, it is not clear how moral agency can be reconciled with determinism, since to be responsible requires that we be free to choose. So there is good reason to be suspicious, if not skeptical, about attempts to relate the sciences to ethics. Either such attempts will be of dubious value because they will run against the facts or they will be downright dangerous because they will, if successful, lead to a diminution of responsibility and of efforts to change things that ought not to be changed. Thus, at first blush, scientific explanations of human actions deriving from biology and psychology seem to be either incompatible with ethical accounts of moral agency or, at best, irrelevant to moral agency. This is the problem of freedom versus determinism.

Second, we have the problem of facts versus values. Even if biology and psychology can describe, explain, and perhaps predict some of our actions, motivations, beliefs, and feelings, it seems doubtful that in using biology and psychology one can ever bridge the apparent gaps between facts and values and between what is the case and what ought to be the case. The descriptions, explanations, and predictions we obtain from these sciences, no matter how good they are, seem to be only about facts. Descriptions of how our brains evolved from those of our hominoid ancestors tell us some interesting things about our history. Descriptions of our neurons and their functioning inform us about some of the current makeup of our brains. Successful predictions that, if we take a certain drug, we will experience great anxiety describe something about a future psychological state. And explanations of that state give us a causal account of how and why the neurotransmitters in our brains, given the effects of the drug we have ingested, bring about the anxiety we feel. So we may be able through the biological and psychological sciences to learn a lot about ourselves, past, present, and future. But what does any of this tell us about what we ought to do? It seems I could have all the knowledge in the world about my brain, its neurons and neurotransmitters, as well as the effects of various drugs on it, and still not know whether taking a certain drug is morally valuable or the morally right thing to do. Facts and values, what is the case and what ought to be the case, seem to be as separate as anything in this world can be. So how can we expect to find significant connections between biology and psychology, which are concerned with facts, and ethics, which is focused on moral values? That is, roughly, the problem of facts versus values.

Finally, there's the problem of information versus meaningfulness. The goals of moral agency lend meaning to our lives. They provide significance to our activities and a sense of direction. Thus they help us to understand and elaborate our potential, actual, and ideal selves in a larger context of persons, society, and the nonhuman environment. For instance, from a Christian religious perspective, our moral life is the way we work out God's plan for our lives. From a nonreligious humanistic perspective, moral agency is the means by which we foster the overall good of the human community, past and present. It is certainly problematic whether biology and psychology can give these sorts of meaning to our moral lives. Evolutionary biology may tell us something about our biological roots. It also tells us that, to some extent, we are the kinds of beings we are today because we have certain characteristics, that is, adaptations, that have enabled us to survive and reproduce more successfully than our hominoid relatives and ancestors who are no longer around. But is that all there is to our lives? So too with what the psychologists might tell us about ourselves. The developmental psychologist tries to describe the stages in human development, from birth to adulthood or from birth to death. Cognitive psychologists tell us how our cognitive capacities work. Social cognitive learning theorists discern the things that motivate us. All this is very interesting. But what does it mean? Why is it significant? It's not clear that the psychologist can answer that question. What we seem to want to know is the meaning of our lives and, perhaps most important, the meaning of ourselves as moral agents. All the biological and psychological information in the world about our moral agency and its sources doesn't seem able to give us that answer. So we have our third problem, the problem of information versus meaning.

But all this is very perplexing. The sciences seem to be the best means to knowledge that we have come up with so far. Why should they pose such problems when we start using them to understand ourselves? Every time conflicts have arisen between scientific findings and what we have taken to be the case on the basis of ordinary understandings, religious faith, or philosophical speculation, it seems science has prevailed. Think of the case of Galileo and the conflict between Copernican heliocentric astronomy and Ptolemean geocentric astronomy. So, too, for human origins. That's not to say that there aren't people, for example, who still hold to a literal Christian biblical account of divine creation. But for anyone who takes the scientific findings seriously, the case for evolutionary origins is very strong. If we are serious about the truth and honest with ourselves, might we not then expect, when and if we have to choose between our ordinary and best scientific accounts of moral agency, that the scientific image of ourselves as moral agents will have to take precedence over our ordinary conceptions?

There are two sorts of general positions about these relationships between the sciences and ethics. First, there are the *separatists,* who argue for a fundamental distinction between the sciences and ethics. If there are any links, the separatists believe they are relatively few and insubstantial. Many religious believers hold that morality is based on theology or sacred books. So, they believe religion and ethics are intimately connected. At the same time, they believe religion and the sciences are fundamentally distinct. For those believers, then, separatism would come naturally. There could be another kind of separatist. Consider a humanist who is a nonbeliever. She doesn't connect religion and ethics, but she might believe ethics derives from experience and disciplined philosophical reflection on that experience. If she also thinks philosophical reflection is independent of the sciences, then she too would probably join the separatist camp.

If you're not a separatist, then you're an *interactionist.* Interactionists contend, in opposition to the separatists, that there are many substantial connections between the sciences and ethics. Among the interactionists, we can discern three types: *antagonists, integrationists,* and *compatibilists.* Antagonists see the relationships between the sciences and ethics to be primarily negative, whereas integrationists view them in a mostly positive light. The compatibilists, on the other hand, although they agree with the other two positions that the connections between the sciences and ethics are many and substantial, believe that in the end the effect of the two on each other is relatively neutral. For instance, consider something like the problem of freedom versus determinism. The separatist might argue that what scientists say about how our actions are determined by causes is one thing and what ethicists say about our moral actions being free actions is another thing. We shouldn't confuse the two by thinking that one contradicts or supports the other. The antagonist believes that freedom and determinism are antithetical and that scientific and moral points of view are incompatible. The compatibilist believes freedom and determinism are compatible and what the sciences might tell us about the determinants of our moral agency does not detract from our freedom. However, the compatibilist does not find that the sciences provide significant support for the existence of the freedom required by moral agency or that they increase one's understanding of it. The integrationists, however, claim that knowledge of biology and psychology supports our belief that, as moral agents, we are free and increases our understanding of what that freedom is and how it is that we are free.

In general, the antagonist believes the same sort of thing about the problems of fact versus value and information versus meaning. The scientist's facts and information have negative import for moral values and meaningfulness. So, whereas the separatist avoids conflicts by not finding significant connections between the sciences and ethics, the antagonist cannot avoid finding significant clashes between the sciences and ethics, since she believes there are significant negative links between the two. The compatibilist, on the other hand, finds no significant conflicts between the sciences and ethics with respect to our three problems, but she contends that overall these disciplines do not contribute significantly to each other's enterprises. The integrationist argues that knowledge of biology and psychology can help to solve the problems of freedom and determinism, fact and value, and information and meaningfulness.

In order to see how an integrationist might address these questions about the relationships between biology and psychology and ethics and handle the problems that attempts to link these areas together seem to raise, we need first to reflect some more on the question of what moral agency is and what a biologically and psychologically based account of moral agency looks like.

1.3   A BIOLOGICALLY AND PSYCHOLOGICALLY BASED MODEL OF MORAL AGENCY

I explore some answers to these questions about the connections between biology and psychology, on the one hand, and ethics, on the other, from the view point of an integrationist's position. Thus I address these issues from a scientific naturalistic perspective. A naturalistic perspective is one that finds morality to be a part of the empirical world and, thus, to be open to understanding through either direct or indirect empirical investigation. What a naturalistic perspective rules out is an exclusive reliance on purely a priori or purely religious approaches, that is, approaches that take moral values out of the material world and place them in some nonmaterial realm, thus making access to them exclusively nonempirical. A scientific naturalistic perspective makes the further assumption that the sciences provide the best empirical knowledge available for understanding morality. But it is not part of this perspective that ordinary perceptual and introspective knowledge are without cognitive merit. Indeed, the scientific naturalist maintains that the sciences are an extension of our ordinary cognitive capacities.

Although all the sciences are candidates for providing relevant input to the enterprise of understanding morality, I focus on biology and psychology. The scientific naturalist does not have to assume that these sciences will, in the long run, make the most important contributions to a scientifically based morality. Indeed, it is quite likely that the social sciences, like sociology, anthropology, political science, and economics, might make much more significant contributions than biology and psychology. But the former are arguably less well developed than the latter, so their potential contributions to an understanding of morality are less clear. But whether this is so or not, the task of exploring the connections between biology and psychology and ethics is sufficiently large and important for study in itself. Nevertheless, it is part of the scientific naturalistic approach I am taking that both the natural and social sciences have a necessary contribution to make to a naturalistic account of morality.

On the scientific naturalistic hypothesis I am pursuing, it is not satisfactory to try to pin down moral agency by merely trying to analyze what is meant by moral agency. Conceptual analysis is an important philosophical tool, but it has its limitations. At its best, it helps to clarify ambiguities, remove inconsistencies, sharpen usage, and suggest vital connections and distinctions. However, I maintain that it is highly problematic that there is some special logic of various types of concepts, discoverable by philosophical skills alone, that determines the usage and applicability of moral concepts, for instance, and to which all empirical research must submit itself. I believe that the history of philosophy and science has shown this view to be fundamentally flawed. Nor should one think that an armchair analysis and discussion of issues of moral agency using ordinary or imaginative examples is enough to capture the phenomenon. Although also helpful, conclusions drawn from the analysis and examination of such cases are likely to reflect the limitations of an individual's experience and a lack of methodological safeguards designed to prevent observational errors and unjustified inferences. Our starting point should make use of what currently seems to best characterize moral agency on the basis of philosophical reflection and ordinary empirical input. What is needed to start with is a working description that does not prejudice significant philosophical or scientific issues relating to moral agency. Ultimately, however, an adequate description is a matter of finding empirical regularities and providing adequate causal accounts of these regularities.

I have said that, on one ordinary view of moral agency, a person acts in a morally correct fashion when she acts on the basis of adequate moral beliefs correctly applied to a particular situation. On this view, then, moral agency seems to require *morally cognitively motivated agency.* As a rough approximation, let's start with the widely discussed two-tier model of human agency to understand moral agency (Frankfurt 1971; Watson 1975; Alston 1977; Taylor 1977; Wren 1982, 1991; Blasi 1983; Baron 1985; Stump 1988). According to that model, agents have at least two motivational systems. The first, a lower-level system, whose basic causal factors are a set of first-level beliefs and desires, has as its function the motivation of behaviors. The second system, a metalevel system, is made up of a set of second-level beliefs and desires about the first-level beliefs and desires and has as its role the influencing of those first-level causal factors. Moral agency is exercised when the determinative second-level beliefs and desires are moral ones.

One of the advantages of this model is that it does not beg any questions about the substantive nature of the moral. That is, it leaves open the question of what content distinguishes moral judgments from other evaluative and normative judgments, such as those concerning social convention, law, religious injunctions, etiquette, or aesthetics. It also makes no determination with respect to whether morality is substantively universal, group centered, or individually centered. Nor does it make any commitments on the balance of benefits to self and others. Another advantage is that it incorporates what seems to be a minimal functional requirement on morality. The functional requirement is roughly pinned down by the condition that the moral agent must be free. The two-level system allows for freedom at least in the soft determinist sense that the agent is not determined in her behaviors by external factors or compelled by first-level psychological factors. However, her second-level beliefs and desires may not be entirely under her control. Using the two-level model, we can say that an agent is acting morally when she acts in accordance with second-level moral beliefs and desires. Moral agency, then, is the capacity to act morally.

One disadvantage of this model for our discussion of the role of biologically based capacities for moral action is that it is a strongly cognitive one, in fact, too much so to be completely satisfactory for our purposes. It is unlikely that primarily evolutionarily based capacities or even all learned capacities, moral or otherwise, are dispositions to believe and desire in any strongly cognitive sense. For example, one may have the tendency to form the belief that it is wrong to inflict pain on an innocent person. Or one may have the desire that this poor child receive some warm clothing. If we understand beliefs and desires to be the objects of propositional attitudes, then the dispositions that enable them are propensities to form propositional attitudes, that is, they are intentional dispositions whose intentional objects are propositions. I am using *intentional* here in the technical philosophical sense of being about something. Mental states such as believing and desiring have the feature of being about something. Neither primarily genetically based dispositions nor dispositions learned by means of conditioning techniques, for instance, seem to be propensities to form propositional attitudes. However, some behavioral dispositions, whether primarily genetically based or learned, do share a characteristic of such intentional dispositions in that they are end-governed. In the case of genetically based dispositions, that end-governedness is due to the fact that the dispositions can be conceived of as biological adaptations, originally developed or currently maintained because they serve ends that promote survival and reproduction. And noncognitive learned dispositions, for instance, those resulting from operant conditioning, are also end-governed insofar as they are the results of previous reinforcing stimuli that are the consequences of the behaviors manifested by the propensity. As we shall see, because of what we know about human development, it seems necessary to postulate such nonstrongly cognitive sources, whether primarily genetically based or learned, for some of our beliefs and desires. For neither phylogenetically nor ontogenetically do we spring into this world as first-level, let alone second-level, agents. Thus I propose to extend the two-level agency model by adding a lower, base level of nonstrongly cognitive motivational level of agency. This new level may consist of either primarily genetically based or learned propensities that have representational and motivational aspects. I leave the nature of these representational and motivational aspects open, except to claim that, insofar as they are not strongly cognitive, they do not take the form of propositional attitudes. These propensities can themselves lead, at times, to behaviors without the operation of the other two levels of the moral agency system. Usually, however, this *base* system, as I call it, works in conjunction with the two-level system.

Consequently, we are led to consider the following three-level system of moral agency. The first level, the *base level,* consists of end-governed propensities to perform certain behaviors. These propensities can be either primarily genetically based or primarily learned. They have representative and motivational features, but they are not fully cognitive in the way that beliefs and desires that are propositional attitudes are. The second-level motivational system, what I call the *behavioral level,* is composed of beliefs and desires about actions to be performed, and it is these that directly bring about action. The second-level system is influenced both by the first-level, base system, and by a third-level system of beliefs and desires, which I call the *reflective level.* This model of moral agency sees moral agency in its full operation as *reflective,* as well as *morally and cognitively motivated agency.* The beliefs and desires of the reflective level among which are moral norms, however these are identified, work directly on the first-level beliefs and desires. Moral action is a feature of the operation of the third-level system, the reflective level. On this model, free actions are those that proceed on the basis of the beliefs and desires of the third-level motivational system. In a free action they determine the beliefs and desires of the behavioral level, as well as the base motivations, that will bring about the action. A relatively maximal condition on freedom would require that the reflective level be operative. A minimal one requires merely that the action performed by base-level or behavioral-level motivational systems be of the sort that a third-level intervention was possible, that is, that they not be compulsive.

I argue in Chapter 6 that a more complete, as well as a more philosophically and empirically adequate account of moral agency requires the postulation of a fourth-level motivational system, a self-system. The self-system allows for self-referential cognition and motivation by including conceptions of the self and motivations arising from such self-conceptions, in particular, conceptions of the self as a moral agent. Given that some of the beliefs and desires of the reflective level concern moral norms, then a fourth-level motivational system is concerned with views of the self in which being a moral person is one of the conceptions of the self that has become highly motivating. This level may, then, influence one to use moral norms rather than other reflective-level considerations in making a decision. In addition, the self-system allows for a greater degree of freedom.

Thus, in the end, I am proposing a model of moral agency that includes four functional levels: (1) a *base level,* constituted by evolutionarily acquired and behaviorally learned capacities and tendencies that incline the agent to act morally in given situations; (2) a *behavioral level,* consisting of a set of moral beliefs and desires that are the immediate sources of actions, and that are influenced by both base-level and higher-level components; (3) a *reflective level,* composed of higher-level beliefs and desires, including moral norms or their equivalents, that influence the behavioral-level beliefs and desires; and (4) a *self-referential level,* containing conceptions of the self, including the self as moral agent, that motivate the use of moral norms and, indirectly, moral action. In the complete model that I am proposing, then, I conceive of moral agency in its fullest extent as cognitively and morally motivated agency that is both reflective and self-referential.[1] However, I do not argue that for an action to be moral, it requires the engagement of all four levels of moral agency.

The biologically and psychologically based account of moral agency I am proposing is based on an integrationist position, one inspired by a scientific naturalistic perspective on the relationships between the sciences and by accounts of moral agency. This scientific naturalistic position comprises six substantive hypotheses concerning the connections between the sciences and ethics.[2] Let's examine these hypotheses as they apply to the relationships between biology and psychology, on the one hand, and an account of moral agency, on the other.

1.4 A SCIENTIFIC NATURALISTIC PERSPECTIVE CONCERNING THE CONNECTIONS BETWEEN BIOLOGY AND PSYCHOLOGY AND MORAL AGENCY

We can distinguish six increasingly controversial, substantive dimensions in which the sciences might make a contribution to ethics and so to our understanding and explanation of moral agency. Specifically, with respect to biology and psychology, I propose the following substantive hypotheses:

1. *Informational Connection Hypothesis:* Because of the relevant information about the circumstances and conditions of moral action that they can provide, the biological and psychological sciences can provide us with factual information important in ethical decision making and for ethical understanding.

An informational connection provides a minimal link between biology and psychology and values. We can suppose that our abilities to act morally — for instance, our abilities to form adequately justified moral beliefs and to act on them — arise independently of genetically and psychologically based capacities and dispositions discovered by the biological and psychological sciences. So, for instance, we know that we have a duty to care for our children, we may justify that duty on the basis, for instance, of our religious beliefs, and we are motivated to carry out our duties because of our religious beliefs; but, nevertheless, we may need to make use of a nurse's biologically based medical knowledge to find out about what sort of diet is best for a sick infant. This need for factual knowledge to aid ethical decision making is generally accepted even by separatists. It provides a minimal connection between the sciences and a noncontroversial one. Thus I shall assume that it has been established and not argue for it.

2. *Explanatory Connection Hypothesis:* Our moral capacities have genetic, developmental, behavioral, and cognitive components. Biology and psychology provide descriptive, explanatory, and predictive knowledge concerning the origin, development, maintenance, use, change, or extinction of the cognitive, affective, and behavioral capacities that are employed in ethical action.

For example, developmental psychology might tell us about the emergence and development of empathy in infants and children. Since the obligation to refrain from harming others is a prototypical kind of ethical duty, we can see how having an empathetic disposition may contribute to the carrying out of this duty. If children have a tendency from their early years to be empathetic, one might argue that such a tendency has a genetic basis. With a typical social and care-giver environment during infancy, we would expect a child to react negatively if another is hurt, to be sad or cry, and spontaneously to seek to help the person who has been hurt. As the child grows, he may learn through reinforcement, modeling, and instruction how to effectively help another person who is in distress. If all this is plausible, and it does seem so, then we would expect biology and psychology to tell us about the evolutionary, developmental, learning, and social cognitive factors that contribute to a person's having and being able to effectively use his empathetic capacities to achieve moral ends.

The existence of these sorts of links between ethics and the biological and psychological sciences is more controversial than the informational connection but still not of a kind that will stir up the opposition of most separatists. Ethical theorists hold that one's obligations are limited in some way by one's capacities. One is not obliged to do what one, through no fault of one's own, is unable to do. Put very generally, human nature sets limits on what our obligations are by placing limits on our capacities. Even separatists are usually willing to admit that biology and psychology, or at least the latter, tell us something about the nature of these capacities, as well as how they come about and are maintained, used, and lost. The kinds of explanatory connections that a model of moral agency postulates are crucial to its ability to handle the problem of freedom versus determinism.

3. *Critical Connection Hypothesis:* An account of the biological and psychological bases of moral agency can effectively critique the claims of common sense or moral theories about human moral capacities and the nature and function of morality.

Assume that sociobiologists are able to demonstrate that there is a genetically based tendency in humans to help relatives, at some cost to the helper,

22

and that social psychologists demonstrate that many people are motivated to help strangers without reward. These findings, based on biology and psychology, could be used to criticize claims that humans are fundamentally selfish and so should not be expected to act altruistically since they cannot. This sort of critical connection between biology and psychology and ethics is clearly a consequence of the explanatory connection. As such, it is as relatively controversial or uncontroversial as that connection. However, in addition to criticisms of ethical claims based on contentions that the view under critique is using an incorrect, incomplete, or inadequate account of human capacities, and so requiring what cannot be delivered, ethical contentions can also be criticized on the basis of normative theories that themselves are maintained to have a basis in biology and/or psychology. This normatively based critical connection is as controversial as the claim that there is a normative connection between the sciences and ethics.

4. *Normative Connection Hypothesis:* From a knowledge of our moral capacities, and on the basis of knowledge from other sources, for instance, perceptual and social, we can formulate some general prima facie normative principles about what it is good for us to do morally and what is morally permissible and impermissible.

So, one might argue, on the basis of what we know from biology and psychology about the human family, that parents have a prima facie obligation to care for their children until they are old enough to provide for themselves. *Prima facie obligation* means that the obligation is presupposed to hold unless there is some good reason why it should not. Of course, we could, with good reason, say that parents realize they have this duty without studying any psychology or biology. And that is correct. But this helps to bring out the point that the sciences are not some esoteric mode of knowledge, completely distinct from ordinary knowledge and reflection on it. To return to our example, what biology and psychology can tell us about the details of our social being and familial ties, as well as about the relative dependence of human infants and children on their parents for development of their human capacities, often, but not always, confirms and refines our experiential knowledge. Nevertheless, this sort of link remains very controversial since the kind of connection it hypothesizes bridges the so-called fact–value gap. Showing how this can be done is a major part of any complete scientific naturalistic account of moral agency. However, I do not attempt to confirm the normative connection hypothesis directly by showing how specific prima facie norms can be supported by findings in biology and psychology. My support for this hypothesis comes indirectly from arguments for the metaethical connection hypothesis, which I formulate as follows:

5. *Metaethical Connection Hypothesis:* If the explanatory hypothesis is substantiated, we can conclude that the capacities of moral agency provide us with relatively reliable mechanisms for achieving moral goals. So we can appeal to their use in attempting to justify moral beliefs, motivations, and actions and in trying to understand the nature and function of morality.

For instance, it might be argued that we can discern from evolutionary biology that morality plays a role in achieving biological fitness and survival. From psychology we might find that morality plays a role in helping us achieve nonbiological ends that we have developed in the course of our individual lives and our social cultural evolution. And if biological fitness and survival or these nonbiological ends are either intrinsically or instrumentally morally valuable, we might be able to justify actions promoting them as themselves morally required or permissible. Just as separatists part company with scientific naturalists with respect to the possibility of normative connections between biology and psychology and ethics, so too do they go their own way with regard to the possibility of a metaethical connection. The satisfactoriness of one's solution to the problem of facts versus values depends significantly on what one says about both the normative and metaethical connections.

6. *Meaningfulness Hypothesis:* An account of the biological and psychological bases of moral agency can make an important contribution to the vision of a meaningful human life, one connected in fulfilling ways to other humans, nonhumans, and the environment.

The vision of the long history of the evolution of life, the richness and variety of living things and of the place of humans in that history, may be a source of meaning and inspiration to human moral agents, giving significance to their attempts to lead moral lives. So, too, may be a vision of an expanding network of humans living in harmony with each other and with nature. The integrationist claims that both biology and psychology have the resources to contribute to the delineation and refinement of these visions and thus to making the life of moral agency meaningful. Separatists, as you no doubt have guessed, demur on this point, too. Thus, what one says about the meaningfulness connection is of prime importance for addressing the problem of information versus meaningfulness.

In broad terms, to be successful, my integrationist account of the biological and psychological bases of moral agency has to provide more satisfactory answers to the questions of moral agency, those of acquisition, action, relevance, and adequacy, than do those approaches that either do not make use of biology and psychology in understanding moral agency or do so to a lesser extent than envisioned in my approach. Although it is essential in assessing a model or theory to examine the relative merits of competitors, my focus is primarily on the positive case that can be made for the integrationist's position. I do not spend as much time showing how the separatist and antagonistic or compatibilist interactionists fare less well in explaining what we know about the intersections of the moral and nonmoral realms as I do on presenting the positive case for my account of moral agency. Insofar as it incorporates the substantive connections I have laid out, the success of the biological and psychological model of moral agency that I am proposing itself provides support for the integrationist perspective that I am advocating. The support for such a perspective will, in turn, recommend the pursuit of the kind of synoptic vision advocated by Sellars and others as a way to pursue and solve philosophical problems by coming to a vision of the whole.

Having laid out a model of moral agency and its integrationist presuppositions, I now think it is time to begin our investigation. In Parts II and III, I focus primarily on parts of the explanatory hypothesis, considering the various ways in which biological and psychological factors contribute to the acquisition and exercise of moral agency. Chapter 2 starts by considering the theoretical bases for claiming that we have evolutionarily based moral capacities. Questions about the existence, nature, acquisition, and function of these evolutionarily based moral capacities will occupy us in Chapter 3.

NOTES

1. Thus far, I have used such terms as *account, theory, hypothesis,* and *model* to characterize my scientific naturalistic proposals about moral agency. I use the term *model* in the phrase *model of moral agency* in a way that is, to my knowledge, common among many scientists, especially biologists and psychologists, who are investigating theories, or parts of theories, or hypotheses. Important discussions concerning the nature of theory have occurred in the philosophy of science literature during the last 30 years or so. Two major competing views of the nature of theory have emerged. The older, logical empiricist conception understands a theory to be a linguistic entity composed of uninterpreted axiomatic calculi in which theoretical terms are given a partial observational interpretation by means of correspondence rules. The more recent semantic account of theories understands a theory to be an extralinguistic entity whose theoretical terms are directly meaningful. In this account, the linguistic formulation of a theory refers to the extralinguistic entity that constitutes the theory.

Moral agency and scientific naturalism

Frederick Suppe (1989) provides a detailed discussion of both accounts, and Beatty (1981), Giere (1983, 1984), Thompson (1985, 1989), and Lloyd (1988) have applied the semantic approach to evolutionary theory. These authors have presented strong reasons for contending that the semantic view of theory structure provides a better understanding of scientific theory generally. Giere (1984) has some especially clear examples of a semantic approach to theories. Beatty (1981), Lloyd (1988), and Thompson (1989) have made very strong cases for the claim that the semantic understanding of theories captures evolutionary theory as it is actually intended and used by working biologists much better than does the syntactic view. Thompson (1989) shows how the semantic view is superior to the syntactic one when it comes to understanding the complex theorizing required in theories like evolutionary theory, which involves a family of theories including theories of evolutionary change, heredity, and genetic mutation. For these and other reasons, especially those concerning the relationships between theoretical and observational terms, I believe that the semantic conception of theories provides a better understanding of theory structure than does the syntactic. Thus, I believe that it better captures the structure of my model or theory of the biological and psychological bases of moral agency. But I do not develop this contention here, and I leave open the possibility of its being interpreted in either a semantic or a syntactic fashion.

2. It may be helpful to compare my approach to the study of moral agency with the traditional conception of the discipline of ethics. Ethics is usually divided into three parts: (a) metaethics, which is about the nature and function of morality; (b) normative ethics, which deals with moral principles and norms; and (c) applied ethics, which is concerned with the application of moral principles and norms to the resolution of moral issues in particular areas of moral concern, for instance, medicine. Since my perspective is theoretical, I am concerned with metaethics rather than normative or applied ethics. As for the six hypothetical connections between the sciences and an account of moral agency, I focus on the explanatory, critical, metaethical (here I use the term in a narrower sense to refer to issues in moral epistemology and ontology) and meaningfulness hypotheses. All these belong to traditional metaethics. One of my primary goals, then, is to construct and support a scientific naturalistic metaethics. In addition to these substantive contributions of the sciences to ethics, specifically to an account of moral agency, there is also the issue of the methodological connections between morality and the sciences. What are the relationships, if any, between ethical and scientific knowing? Scientific knowing is typically described in terms of observation, experimentation, the formulation and testing of hypotheses and theories, prediction, and explanation. The issue of methodological connections concerns the question of whether there are similar sorts of cognitive processes involved in ethical knowledge. Although I believe that there are significant methodological connections between a scientific naturalistic account of ethical methodology and the methodologies of the sciences, I cannot pursue that issue here.

II

The biological bases of moral agency

2

Evolution and morality
Can evolution endow us with moral capacities?

In this chapter and the next, I examine the question of whether humans possess an evolutionarily based moral agency. Assuming that humans are moral agents, I seek to establish that there are biologically based reasons for postulating that humans possess evolutionarily based moral tendencies and capacities as part of a base level of moral agency. In these two chapters, I lay out the grounds for the theoretical plausibility of the claim that humans possess such capacities and tendencies. In Chapter 4, I argue for the existence of one such capacity, the capacity for empathetic distress. Even if these arguments are convincing, their limitations should be obvious. One biologically based moral capacity doth not a base level of moral agency make! However, I think that it is plausible that there are other base-level, evolutionarily founded moral capacities; for instance, for parental care. To develop a reasonable case for the existence of some or all of these moral capacities would be a daunting task. My hope is that I will make a reasonably good case for one such capacity and that, in so doing, I will illustrate the plausibility and the potential fruitfulness of my approach and model.

To establish the theoretical plausibility of evolutionarily based moral capacities I discuss evolutionary theory, since it provides the fundamental basis for biological altruism (Section 2.1), and then present some of the results of sociobiology, because it serves as the proximate theoretical source for an account of biological altruism (Section 2.2). I then begin to lay out the links between biological altruism and morality by first examining the relationships between biologically based altruism and ordinary altruism (Section 2.3). Since the theoretical plausibility of an evolutionarily based moral agency depends crucially on one's account of what counts as moral, I discuss in Section 2.4 the question of what makes something a moral phenomenon. I then lay out in Section 2.5 a theoretically plausible story about the phylogenesis of evolutionarily based moral capacities.

Biological bases of moral agency

2.1 EVOLUTIONARY THEORY

First, we must distinguish the *fact* of evolution – modification with descent, as Darwin called it – from *explanations* of evolution. Evolution can be described in terms of the changes in organisms over time that have led to the diversity of past and present life forms. These changes are reflected in such phenomena as biogeographical distribution, the order of the fossil record, the morphological and physiological similarities between past and contemporary forms, resemblances among contemporary organisms, and the adaptations of organisms to their environments (Darwin 1859; Kitcher 1985; Ruse 1986). Often, contemporary biologists define evolution in terms of any change from one generation to another in the units of heredity, the genes (Futuyma 1979, p. 7). One of Darwin's major contributions is that he offered a powerful way to explain these evolutionary phenomena: his theory of natural selection. He also suggested two other mechanisms: sexual selection and the Lamarckian mechanism of the inheritance of acquired characteristics.

Theories provide explanations of events or laws (Kitcher and Salmon 1989). We can take theories as answers to questions about how or why something happened. By answering these questions about evolution, we believe that we have found out some of the causes of evolution. The theory of natural selection is offered as an explanation of the facts of evolution. So we can think of the theory of natural selection as explaining the facts of evolution by offering an answer to the questions of how and why life forms are distributed the way they are geographically, the fossil record is as it is, life forms have the morphology and physiology they do, current and past life forms are alike and different, and so on. The theory of natural selection, when applied to particular questions about specific evolutionary phenomena, provides us with knowledge of the causes of these phenomena or the mechanisms by means of which they occurred.

The theory of natural selection is a type of selection theory (Darden and Cain 1989). In other words, it is one of a class of theories that has a distinctive explanatory role. Selection theories are distinctive because they answer certain sorts of questions, those that are concerned with the adaptation problem. In general, we can say that the adaptation problem is a problem of how and why varying elements in an environment come to fit and do fit that environment more or less well. The form of the adaptation in question, and thus of the type of selection, depends to a large extent on the types of elements and environments under discussion. What do selection theories look like? Consider a classical example of natural selection: the case of the peppered moths (Kettewell 1973). The phenomenon to be explained is the change in frequency of the number of speckled peppered moths relative to the black-colored variants in urbanized, industrial areas in England. It was noticed that as cities became industrialized, the melanic forms became much more common in the industrialized areas, whereas the speckled form continued to be predominant in the nonindustrialized countryside. Kettewell used the theory of natural selection to explain this phenomenon and suggested that camouflage was the central adaptive feature that was selected for by the environment. He hypothesized that the peppered moths were better hidden from their predators on trees in nonpolluted areas because these trees had lichens growing on their trunks that made them almost invisible to their predators. On the other hand, the melanic form of the moth stood out on the lichen-covered trees. But with the increase in pollution in the urban areas, the lichen was killed, and the trunks of the trees become dark colored. Thus the speckled moths, rather than the melanic moths, were much more noticeable by predators in the new situation. As a result, the population of speckled moths declined in the urban areas where pollution killed off the lichen on the trees. There are, of course, other selection accounts of how the evolutionary phenomenon might have occurred, that is, selection for adaptations other than coloration. But Kettewell ruled these out as less likely. So his Darwinian story is well confirmed if it can explain the data better than its alternatives, as well as account for data that the alternatives cannot explain.

Thus, by means of a selection theory, we can answer the questions of how the adaptation occurs and why. Of course, we did not answer all the "how" and "why" questions that arise in investigating the phenomenon. For instance, we might want to know what genes are associated with coloration and how they produced the coloration. Or we might want to know what other functions, if any, are served by coloration. I shall have more to say later about why questions and their answers in this and other kinds of selection explanations; but it should be clear that in this case, that to answer the question about why the color camouflage occurs in terms of its effects — namely that it increases the chances of survival and reproduction of the moths that possess it — is to give an explanation in terms of the consequences of the property, the adaptation, in question. Indeed, the property is an evolutionary adaptation precisely because it is a heritable property that had that sort of consequence in the evolutionary history of the organism (Godfrey-Smith 1994). Alternatively, to answer the question of *how* the distribution of colored moths has occurred is to appeal to the key property of the moths in question and to the critical factors of the environment, that is, the coloration of the trees and the visual discriminatory capacities of the birds that are the moths' predators. We explain how the results of the

31

interaction occurred in terms of these key antecedent characteristics of the moth, the birds, and the trees.

Consequentialist explanations have been called *teleonomic* to distinguish them from *teleological* explanations, especially purposive explanations (Mayr 1974). In other words, Darwin, by explaining biological adaptations in terms of natural selection rather than immediate divine creation, did not substitute the intentions of nature for those of the divinity. On the other hand, explanations of an organism's characteristics in terms of their fitness-enhancing properties differ from so-called mechanistic explanations precisely because they appeal to the consequences of the phenomenon to be explained rather than to its antecedents. In this respect, they are like functional explanations. We can explain the presence of the light switch in terms of what it does, that is, its turning on the lights in the room. And we can explain the presence of the lungs in an animal in terms of what they do, that is, enable the intake of oxygen into the bloodstream.

One further thing should be noticed about the preceding characterization of the theory of natural selection. Implicit in this is the key idea that in natural selection the property in question is heritable. Roughly, what is meant is that the characteristic is passed on from generation to generation by means of factors that are transferred in reproduction. These factors, of course, are what we have come to identify as genes.[1] The developmental process by which the genes associated with the phenotypic property bring that property about, in our case the process by which they bring about the coloration of the moth, is a complicated one involving not only the action of the genes but also the environments in which the moth develops. A reason why it is very important to recognize explicitly that the property selected for is a heritable one is that other types of selection – for instance, behavioral, cognitive, and cultural – can mimic natural selection. Nonheritable variants – for instance, learned behavioral or cognitive tendencies – can be selected for either because they are biologically more fit than their competitors, that is, they lead to differential survival and reproduction, or because they are better fit to produce some nonevolutionary consequence for which the given environment is exerting a selection pressure.

The traits selected for in natural selection benefit the organism in the sense that they increase the relative fitness of the organism and, thus, its chances for survival and reproduction. It is not immediately apparent how altruistic characteristics, those that do not benefit the organism, can arise by means of natural selection. Yet organisms seem to display altruistic characteristics. How can this be? Let's call this the *problem of altruism*.[2] The problem of altruism is of particular interest to us since altruism is often associated with morality. To see how biologists have addressed this problem, we now take a look at the biological subdiscipline of sociobiology.

2.2 SOCIOBIOLOGY

Sociobiology brings together a number of biological disciplines in the study of the origin, development, maintenance, and change of animals', including humans', social behaviors. These disciplines include evolutionary theory, population genetics, neurophysiology, developmental biology, ethology, and comparative psychology. At its heart are theoretical developments in evolutionary theory and population genetics. Although we do not go into the technical details of these developments, we need to see how sociobiology has extended the potential explanatory power of evolutionary theory. It is this extension that is the source of the gleam in the eye of sociobiologists like Wilson and Alexander, who imagine and propose important links between biology and ethics. So, first to the reasons for the development of sociobiology and then to the possible connections between sociobiology and ethics.

Biological organisms often interact with each other. We think immediately of mating; parental care; parent–offspring interactions; sibling interactions; predator–prey interactions; intraspecies competition for territory, mates, and resources; various hierarchies of relationships; and intraspecies cooperation. Many species are social in nature: They live in groups and have intricate social arrangements. Can these phenomena, their evolutionary origin and maintenance, be explained by the theory of natural selection? Since some of the traits that make for these social behaviors of organisms seem to be designed to promote the survival and reproduction of the individuals who possess them, it looks like prima facie they are candidates for selection explanations. But certain sorts of altruistic behaviors appear anomalous, for instance, behavioral tendencies to emit alarm calls, which appear to be tendencies to act altruistically toward others who are not directly related to one in situations where it seems that there may be no opportunity for reciprocation. It is not clear how such traits can be shown to promote the survival and reproduction of those who possess them. Thus the problem of altruism.

Biological altruism is concerned with evolutionarily based behavioral tendencies that lead to behaviors that enhance the fitness of the recipient of an action at the expense of the donor. Fitness effects are measured in terms of their contribution to differential survival and reproduction. If nature

33

selects for those organisms that are more fit than their competitors, then it seems that nature would select against organisms that behave in ways that benefit others over themselves. Darwinian natural selection should select for selfish traits, not altruistic ones. So it seems.

To help clarify the problem of altruism, let's make some preliminary distinctions between biological altruism, psychological altruism, and morality. First, although all forms of altruism have to do with providing benefits to another at some cost to oneself, we can distinguish them on the basis of the types of benefits involved in each case. Biological altruism is concerned with fitness benefits, which redound to survival and reproduction; psychological altruism and moral altruism are not so confined. Thus, for example, suppose that I give you a painting by Renoir out of the goodness of my heart. On the other hand, suppose that I give you that same painting so that you can raise money for the poor or so that you can hide it from the police, who are hunting for it because it was recently stolen from the local museum. Second, psychological altruism and morality necessarily involve cognitive mechanisms, such as beliefs, motivations, and intentions, as proximate sources of behavior; biological altruism, however, need not invoke such psychological mechanisms and may be concerned only with behavioral tendencies or noncognitive mechanisms. To put the distinctions in other terms, biological altruism is not necessarily connected either substantively or functionally with psychological altruism or morality; that, of course, is not to say that it is necessarily unrelated either. How then does the neo-Darwinian explain the origin and maintenance of biological altruism?

One straightforward explanation is that organisms act for the survival of the species. If, of course, they did, then there would be no problem in explaining altruistic behavior in organisms, that is, behavior that benefits some other individual organism of the same species. Nature would be selecting for those traits that benefit the survival of the species. But the facts don't support this interpretation of natural selection. Intraspecies competition for resources and intraspecies killing, as well as cooperation and altruism that is limited to groups much smaller than the entire species, all show that this understanding of what nature selects for is incorrect. On the other hand, the problem of altruism does not arise because nature selects for the survival of the individual. Survival of the fittest does *not* mean the survival of the *individual.* Darwinian natural selection is concerned with the *survival* of the individual organism only insofar as it is connected with *differential reproduction* by individuals. Differential fitness is measured ultimately in terms of differential reproduction. So, according to this understanding of Darwinian natural selection, nature selects for adaptations that promote differential reproduction. Individual selection is concerned primarily with the direct descendants of individual organisms. Thus, according to Darwinian theory, a purely selfish individual, one concerned only with his own survival to the exclusion of the survival of anyone else, is *not* the fittest because such an individual does not survive to reproduce, since that involves, at a minimum, child bearing and rearing, behaviors that benefit others, one's children, at some cost to oneself. Often such acts are not reciprocated. But even if they are reciprocated so that the costs are made up, unless the benefits lead to increases in fitness, the theory of natural selection cannot explain them, except indirectly. That is, such behaviors and the behavioral tendencies that produce them will be selected because they are connected with some other traits that do promote differential survival and reproduction. Consequently, neo-Darwinian theory can explain certain types of behaviors considered altruistic in the ordinary sense of the term, such as helping one's children, since that benefits individual fitness, and helping related and nonrelated others who reciprocate, since that also benefits individual fitness. But what neo-Darwinian theory seems unable to explain is unreciprocated altruistic behavior toward others who are not one's direct descendants, either kin or nonkin. Genes that promoted traits for that kind of altruism would, it seems, be weeded out of a population. If there is no such thing as species selection and if individual selection cannot explain such altruistic behavior, what can? Is Darwinian theory incomplete or even incorrect?

One plausible way to show how Darwinian theory could explain such altruistic behavior is called *group selection*. A group is usually considered to be a population that is smaller than the entire species but that includes more than the organism and its direct and indirect descendants. Group selection might occur if altruism promotes the good of the group rather than that of the individual. Groups of altruists may have a competitive advantage over groups of selfish individuals. For instance, we might think that organisms regulate their clutch size in order to maintain an optimum population size for a given environment. Consequently, their behavior benefits the group by reducing the likelihood of an overuse of resources leading to the demise of the entire population. Two sorts of problems have been posed for this scenario (Williams 1966). First, an alternative account in terms of individual selection may be available. Thus, in our example, regulation of clutch size might occur because higher population density makes fewer resources available and, thus, may lead to less reproductive activity or less successful reproductive activity. Alternatively, the phenomenon can be explained in terms of individual adaptations that are selected for because they benefit the individual and, as an accidental consequence, benefit the population. Second, it appears that from a theoretical perspective, group selection, if it occurs, will not be a very widespread phenomenon. The apparent theoretical problem with group selection is that within groups selfish individuals always seem to have the advantage over altruists. Since selfish individuals, who seek their own survival and reproduction, have the advantage in encounters with altruists, who seek to benefit the group, the former will tend to outperform the latter and eliminate them from the population. The group will be subverted from within. Thus, for traits that incline an organism to behave in ways that benefit the group rather than the individual to persist in a population, conditions have to be of the right sort. One such condition is that altruistic groups will have to divide or split up, that is, reproduce at a higher rate than it takes the group to become selfish by subversion from within.

Another way in which altruism that is beneficial to the group might maintain itself is if the benefits accruing to the altruistic individuals in their interactions with each other offset the costs they incur in encounters with selfish individuals so that they are not at a reproductive disadvantage with their selfish brethren. Since the time of Williams's (1966) penetrating critique of group selection, and until relatively recently, many, if not most, evolutionary biologists deemed group selection to be a theoretically possible but actually rare phenomenon. But things seem now to be changing. Clarification of the nature of selection, group selection, and altruism, as well as, arguably, the identification of more instances of group selection seems to be changing the picture for both group selection and altruism directed toward the benefit of the group. (Sober 1984; D. S. Wilson and Sober, 1994a,b).

But there is a less controversial path to the explanation of altruism, one that is intimately associated with the development of sociobiology. One clue to the solution of the problem of altruism is that nonreciprocated altruistic behavior is most often directed toward either direct descendants or indirect descendants, kin. We have seen that Darwinian individual natural selection explains nonreciprocated altruistic behavior toward direct descendants. So our problem focuses on nonreciprocated altruistic behavior toward kin. The sociobiological explanation of kin altruism, developed largely by William D. Hamilton in the 1960s (Hamilton 1964a,b; Kitcher 1985), accounts for such altruism by showing how it promotes the differential survival and reproduction of the gene line, that is, not merely of the direct descendants of an individual but also of its nondirectly related descendants. The claim is that natural selection operates in such a way that those traits, including behavioral tendencies, that increase not only the fitness of the individual but also that of its kin, are selected for. As a result, although kin altruistic behaviors might decrease individual fitness, they could increase inclusive fitness. Leaving out some complications, inclusive fitness means, roughly, fitness measured not only by direct descendants but also by indirect descendants (Kitcher 1985, p. 83). Altruistic behaviors toward kin may decrease the likelihood of reproducing directly but increase the likelihood of reproducing indirectly through one's kin, for example, through one's siblings. Thus, genes promoting such kin altruistic behaviors could be passed on and would not necessarily be weeded out. It turns out that one can, as Hamilton did, make precise calculations in terms of reproductive benefits about when to expect kin altruistic behaviors. This sociobiological explanation of kin altruistic behaviors in terms of kin selection and inclusive fitness is a natural extension of Darwinian explanations in terms of individual selection and fitness.

Thus we see that Darwinian natural selection, now expanded by means of the concept of inclusive fitness to include both individual and kin selection, can in theory explain nonreciprocated altruistic behavior toward direct and indirect descendants and reciprocated altruistic behavior toward nonrelatives. Whether in fact it does explain such behavior, and the extent to which it explains such behavior, is, of course, another matter. But the expanded theory does entail and predict such behavior. Of course, common sense and other psychological and social theories do too, at least in the case of humans, but on very different bases. So the sociobiological extension of neo-Darwinism is, in the first instance, only a theoretical advance: It shows how the phenomenon of nonreciprocated kin altruistic behavior is a consequence of and is compatible with neo-Darwinism. But is the sociobiological explanation of altruism the best or only explanation of such behavior? That is a question about the empirical success of this kind of explanation, and it is quite a different sort of question, one that we address later. We are ready now to return to moral agency and the connection of our discussion of evolutionary theory and sociobiology with questions of moral agency. Our first step in that direction will be to see how ordinary and biological altruism are related.

2.3 BIOLOGICAL AND ORDINARY ALTRUISM

To facilitate our discussion of the relationships between neo-Darwinism as supplemented by sociobiology and moral agency, let's first summarize and put together some things about altruism. We can distinguish our ordinary understanding of altruism and selfishness from the biological meanings of altruism and selfishness, that is, the technical meanings of altruism and selfishness deriving from biology. When we do that, we get the following:

1. *Ordinary altruistic behavior:* doing something to benefit another, with the motivation of helping another in preference to oneself.
2. *Ordinary selfish behavior*
   a. *Straightforward selfishness:* doing something to benefit oneself, with the motivation of helping oneself, or
   b. *Enlightened selfishness:* doing something to benefit another, with the motivation of helping oneself.
3. *Biologically selfish behavior*
   a. *Directly selfish behavior:* (i) doing something that has net benefits for oneself, and thereby one's children, with the consequence of increasing one's individual fitness or (ii) doing something that has net benefit for one's children, with the consequence of increasing one's individual fitness.
   b. *Reciprocally altruistic behavior:* doing something that has net benefits for someone who is neither a direct descendant nor a kin, and thereby either increasing one's individual or inclusive fitness because of the reciprocation of the benefit by the recipient (or someone else) to oneself, a direct descendant, or one's kin.
4. *Biologically altruistic behavior*
   a. *Kin altruistic behavior:* doing something that has net benefits for one's kin (nondirect descendants) and thereby increasing one's inclusive fitness.
   b. *Group altruistic behavior:* doing something that has net benefits for the other members of a group (not necessarily directly or indirectly related others) and thereby increasing the group's fitness.

What are the relationships between ordinary and biological selfishness? First, note that our descriptions of selfishness do not include what we might call *definitional selfishness,* that is, any behavior of an agent is by definition selfish because it is a behavior of that agent. As an extension of this notion of definitional selfishness, one might say that any behavior an agent wants to do is, by that very fact of the agent wanting to do it, a selfish behavior. We do not include these definitional forms of selfish behavior under either the ordinary or biological senses because they do not help us distinguish altruistic from selfish behaviors. If we used definitional selfishness, then both ordinary and biologically altruistic acts would be selfish, and we could not distinguish them from ordinary and biologically *selfish* behaviors without distinguishing between different types of definitional selfishness. That just creates extra work, something we surely don't need.

How do biological and ordinary selfishness and altruism differ? First, there is the distinction that we made earlier between types of benefits.

Biological altruism and selfishness are limited to benefits connected with survival and reproductive success; ordinary altruism and selfishness have no such limitation on the nature of the benefit. It could be the case, although it seems unlikely, that all benefits are connected with survival and reproduction, either positively or negatively; but that's an empirical issue that needs to be settled empirically. On the other hand, some of the genetically based characteristics may be neutral or relatively insignificant with respect to differential survival and reproduction. Second, there are distinctions concerning the recipients of the benefits. Fitness benefits are not benefits for the organism itself or for its genotype since, as we have seen, fitness is measured ultimately in terms of reproductive success. Nor are they even benefits for the organism's genes, since these genes will perish with the organism. Rather, they are benefits for the copies of the organism's genes, for gene types rather than gene tokens. On the other hand, ordinary altruism is concerned with benefits to the recipient of the altruistic behavior, and ordinary selfishness with benefits to the agent herself in the case of selfish behavior, not necessarily to her genes or genotype or to the genes or genotype of the recipient of the benefits. Third, there is a distinction with respect to behaviors and motivations. Biological altruism and selfishness need only be concerned with behaviors and their effects. Thus evolutionarily selfish behaviors, for instance, caring for children, could be carried out by psychological mechanisms that have either a basically altruistic or selfish motivational structure, and evolutionarily altruistic behaviors, for instance, behaviors benefiting the group at a cost to oneself, could be brought about by psychological mechanisms that have either a basically selfish or altruistic motivational structure (Sober, 1989a,b, 1993, 1994a,b). Nor is it necessary, as we shall see shortly, that nature select only for motivational systems that are basically selfish, although, of course, the measure of the success of the adaptation would be the behavioral contribution to individual, inclusive, or group fitness. On the other hand, ordinary selfishness and altruism include motivations.[3] They often include knowledge and freedom if we are talking about humans. That is not to say that the ordinary terms are not sometimes applied to actions merely in terms of their effects. Nor is it to say that the application of the ordinary terms requires cases in which maximum knowledge and freedom are evident. There is a fair amount of flexibility in ordinary usage, but, as a rule, ordinary usage includes motivations. Thus, although both biological and ordinary uses of *altruism* and *selfishness* exclude a mere accidental connection between a behavior and the one who benefits, ordinary usage can include a continuum of motivational modes ranging from nonconscious mechanisms through emotionally motivated acts to planned acts with intentions for short- or longer-term effects.

39

Given these major differences, we can see immediately that biological selfishness *excludes* ordinary selfishness if the benefits of that selfishness have no connection with survival and reproduction. To put it another way, nature selects against those individuals who have genes that promote behaviors that benefit the individual organism exclusively, and not its direct or indirect descendants. Whether there are such genes or not is another question. Nevertheless, it is important to understand that the *selfishness,* as well as the *altruism,* promoted by natural selection falls into the category of ordinary altruism only insofar as its behavioral effects are considered. Leaving aside the question of motivations for the time being, we can make the following comparisons: Helping oneself, as opposed to helping one's children, kin, or nonrelated others, is both biologically and ordinarily selfish. Helping one's children is biologically selfish but ordinarily altruistic. Helping one's kin is both biologically and ordinarily altruistic. If reciprocation is present, helping nonrelated others is biologically selfish but ordinarily altruistic. Helping nonrelated others, without reciprocation, is biologically and ordinarily altruistic. Bringing in motivations, ordinary altruism and selfishness are distinct from either biological selfishness or altruism that is brought about by nonmotivational mechanisms since the former includes some motivations. However, if nature also selects for motivational mechanisms to bring about its selective ends, then ordinary and biological notions of selfishness and altruism overlap with respect to their proximate modes of activation.

Granted these relationships between biological and ordinary selfishness and altruism, how, if at all, do biological and ordinary altruism, especially the former, fit in with morality? More specifically, what makes something morally relevant, and is biological altruism morally relevant?

2.4 WHAT MAKES SOMETHING A PART OF THE MORAL REALM?

What are moral phenomena? There is a range of what we might call moral phenomena, all of which, hypothetically at least, could be related to the sciences. In the most general terms, ethicists are interested in the morally good and bad, what ought to be or ought not to be, what is morally right and wrong, what is morally permissible or impermissible. Phenomena that are characterized morally include cognitive, motivational, and behavioral capacities and dispositions, as well as feelings, evaluations, judgments, intentions, decisions, actions, and their consequences. They also include habits, virtues, character, professions, and ways of life. Moreover, individuals, groups, institutions, societies, nations, groups of nations, historical periods and eras, and so on might all be characterized morally. We can have intuitions, feelings, perceptions, concepts, generalizations, norms, principles, and theories about morality, as well as views on the moral meaning of life. Indeed, academic disciplines like philosophical and religious ethics are devoted to the study of morality.

Our central question concerns the phenomenon of agency. Specifically, we are interested in the relationships of our biological and psychological capacities and tendencies with our moral agency. Agency clearly is a complex phenomenon that includes many of the phenomena that we have mentioned as potentially moral phenomena. We are able to be moral agents because we have certain cognitive, affective, and behavioral capacities. By our agency we develop habits, virtues, and a character that can be described, explained, understood, and evaluated in a variety of ethical ways. Recall that on the ordinary view of moral agency that we discussed in Chapter 1, a person acts in a morally correct fashion when she acts on the basis of adequate moral beliefs correctly applied to a particular situation. But we still have not gotten to the question of what morality is or what it is to act morally or be moral. All we have done is to talk about the phenomena that are usually described in those terms. It's time to bite the bullet.

We are asking for the criteria that make something morally relevant. The relevance criteria of moral agency are those of its features that mark it off – if it is to be so marked off – as belonging to the realm of morality as opposed to amoral or nonmoral kinds of agency, for instance, those of etiquette, prudence, personal preferences, social custom, social role, religious custom, technical competence, the law, and aesthetics. *Substantive* relevance criteria answer questions concerning who are moral agents and those worthy of moral concern, as well as questions about the nature of the actions being performed. They concern the content of both the actions and intentions of the agent. *Functional* relevance criteria answer questions concerning the kind of agency that is required for moral agency; they concern the form of the actions and intentions of the agent.

Substantively, morality can concern both other-regarding and self-regarding actions and intentions of an agent. This is not to say, as we indicated earlier, that other aspects of persons, their habits or character, for instance, and other entities, groups of people or institutions, for instance, cannot be characterized morally. Other persons can include one's children, kin, friends, strangers who belong to one's group, strangers who do not, members of one's ethnic group, a nation, or all people. Some would include nonhuman animals or even all of nature on this list. They may be right to do so; but not doing so, at this point, begs no questions. Actions can benefit or harm and intentions can be aimed at benefiting or harming. Moral actions and intentions can be discussed either in terms of providing benefits or in terms of refraining from harming. I shall use "providing benefits" in an inclusive way to include omission of harmful acts. Moral actions and intentions that are other-regarding can benefit, or seek to benefit, the other only, the other more than oneself, the other equally with oneself, or the other in some degree but less than oneself. A highly inclusive set of substantive criteria would include self-regarding as well as other-regarding actions and intentions and would require in the latter only that the intention be to benefit the other in some degree. A maximally restricted set would include only other-regarding actions with only the intention to benefit the other, not the self.

Of course, to determine the nature of these benefits and harms is a major substantive issue. Experience gives us some clue as to what sorts of actions might generally be beneficial or harmful. We could probably all agree that the provision of certain sorts of material, physical, psychological, and social help would fall under benefits and the failure to provide such or the deprivation of such would fall under the category of harms. Ethical theories, of course, have attempted to describe, understand, and establish what these benefits and harms are. Utilitarians speak of pleasure and pain or happiness and unhappiness. Deontologists fix on respect for the person. Naturalists and others believe that we can discover what is basically beneficial and harmful to humans by studying human nature and determining the sorts of things that fulfill human capacities and needs. Theological ethicists bring in a reference to the divine will or to cosmic purposes. Besides using experiential, philosophical, and theological approaches to fixing on the content of morality, we can examine the kinds of actions that seem to be of moral relevance to scientific investigators, such as social psychologists and sociobiologists who take an empirical approach to the issue. Very often we find social psychologists investigating altruistic (or what they often call *prosocial*) behavior, cooperation, aggression, adherence to norms, and moral decision making and reasoning. Sociobiologists draw a link between altruism, cooperation, and morality.

There are some sharp disagreements about what should be considered substantively morally relevant. Are hygienic and dietary practices within the moral realm? Do economic policies and organizational practices belong in morality? How about modes of social address or laws about zoning or property rights? We cannot answer these questions. Nor can we determine whether the experiential, philosophical, theological, or empirical approaches have best defined the realm of the moral. Yet, to proceed, we need to make some decisions and hypotheses about what substantively belongs to the realm of the moral. This is not an unusual position for a scientist or, indeed, any investigator to be in. One must start somewhere, with some description of the phenomenon one is investigating and attempting to adequately describe, explain, and understand. But one also realizes that initial descriptions may have to be revised as the investigation proceeds. Issues of substantive relevance will be important for us when discussing the possible contributions of both sociobiology and psychology to ethics. We shall see that, for the most part, psychologists have been more attuned to substantive relevance issues than sociobiologists.

Freedom and sufficient knowledge are commonly acknowledged as functional criteria of moral agency. *Freedom* here refers to freedom of choice. Negatively, free choice involves a denial of all or some determinism. At this point, we can note what seem to be the two extremes of the continuum. The first, on the end of maximal freedom, often called *libertarian freedom,* posits the ability to choose without determination. Of course, the free choice itself is determinative of the action chosen; but the free choice is not itself determined. An agent, when free, performs her action by choosing freely. On the other end of the continuum is the minimalist position, the *soft determinist position.* According to soft determinists, free choice is not something that occurs in the absence of all determination, but only in the absence of some sorts of determination, for instance, physical coercion. Psychologically compelled or physically coerced behavior falls outside the continuum of functional moral relevance. These sorts of behaviors would be determined in such a way that they would be nonfree. So these behaviors would be neither moral nor immoral, but amoral.

To be sufficient, knowledge must be about both what is being done and the relevant norms. Moral credit or blame requires that an action's motivation make a connection with relevant norms. Relatively maximal functional requirements for moral credit restrict moral agency to those actions in which moral norms are either explicitly or implicitly invoked. Less maximal requirements would include actions performed habitually because of previous invocations of norms. For moral blame, the norms should and could have been invoked but were not, or were invoked but were not followed. Relatively minimal functional requirements would eliminate the requirement that moral norms or principles be operative or reinterpret their nature in a minimalist fashion. For instance, the requisite moral knowledge may refer to hypothetical rules rather than categorical ones. Or it may refer to certain recipes about behavior, commands, paradigm cases, or prototypes, or may be even lists of moral or immoral behaviors.

Given the ambiguities about the nature of morality and the widely different accounts of its nature, it is best, I think, to adopt a cautious approach to the setting down of relevance criteria. My general strategy with respect to both substantive and functional relevance criteria will be (1) to try to choose cases and use descriptions that are paradigmatic and seem to have the most agreement and (2) to try to avoid begging any crucial questions. Consequently, I understand actions to be functionally morally relevant that are performed on the basis of what corresponds to the operation of the reflective level of agency in the model I am proposing. Although I have made the use of norms constitutive of that level, I allow a wide interpretation of the nature of norms so as to include the wide range of possibilities mentioned earlier. In addition, I allow actions to meet the functional relevance criteria, even if the reflective level is not put into operation, as long as it could have been activated. I do this to allow for skilled moral action that sometimes does not require the operation of higher-level components of agency, although it remains open to their invocation and, indeed, may have been fashioned by their previous use. With respect to substantive moral relevance, I do not insist on the inclusion of actions benefiting the self, and with regard to those benefiting others, I take as paradigm cases actions that benefit the other more than the self or that, in situations of choice about the balance of benefits going to self or another, favor another rather than the self. I also think that it is important to recognize that the description of the explananda themselves, that is, moral actions and moral agency, are themselves importantly related to theoretical choices and potential theoretical explanations. Thus these descriptions may need to be revised more or less radically in the course of investigation.

Now that we have some sense of what counts as an evolutionarily based capacity and the connections between biological and ordinary altruism, as well as their connections with morality, we can address directly the question of whether evolution can endow us with moral capacities. Using either the ordinary criteria of moral agency, discussed briefly in Chapter 1, or the minimal ones just examined, we are moral agents. Some of the morally relevant actions we perform concern having and rearing children, interacting with relatives, and interacting in various ways with individuals with whom we are not directly related. Part of being able to perform these actions is that we have certain tendencies that in various situations incline us to perform these actions. On reflection or without it, we do or do not act on these tendencies. The question I am asking about the possibility of evolutionarily based moral capacities concerns whether some of these already identified tendencies that function, at times at least, as a component of the more complex sort of agency that we exercise could themselves have an evolutionary origin. The sense of possibility involved here is stronger than that of logical possibility. I assume that the claim that some of these ten-

44

dencies and capacities are evolutionarily based is not logically contradictory. The sort of possibility I have in mind concerns evolutionary possibilities. Given what we know about evolutionary theory, sociobiology, and other relevant biological disciplines, does it follow that humans *could* have evolutionarily based moral capacities?

2.5   THE POSSIBILITY OF EVOLUTIONARILY BASED MORAL CAPACITIES: A PLAUSIBLE PHYLOGENETIC STORY

One way to give an affirmative answer to our question is to weave an evolutionary story showing how nature could have selected for such capacities in our evolutionary history. Thus it may be helpful, before we move from questions about theoretical possibilities to empirical actualities, if we conclude by spelling out a theoretically plausible story of the phylogenetic origins of these moral capacities, in particular, of those altruistic capacities that we have identified as prima facie substantively morally relevant. But we need to be forewarned that theoretically plausible stories should not be turned into adaptationist likely stories (Gould and Lewontin 1979). Theoretical accounts become legitimately likely on the basis of empirical evidence; but as yet, we have not considered in any systematic way the evidence for the existence of evolutionarily based moral capacities. That task is still to come. Adaptationism is a variously construed failing of those who employ explanations of traits as adaptations, that is, as characteristics that have been selected for because they promote differential fitness for their bearers. One way to construe adaptationism is as a sin of excess. Adaptationists make all or too many biological traits adaptations. As a result, natural selection becomes the only, or the predominant, evolutionary force. However, some traits may be free riders, objects of selection, not because they provide differential fitness advantage to their bearer but because they come along with other traits for which there is selection. In addition, we have seen that evolutionary biologists postulate evolutionary forces other than natural selection. Moreover, in the case of moral agency, we need to be very open to the likelihood that a substantial proportion of its component capacities are molded by nonevolutionary factors. Determining whether a trait is selected for is an empirical matter, as are judgments about the extent of operation of natural selection in our phylogenetic history. Adaptationism is best avoided by careful distinction between theoretically plausible accounts of phylogenetic history and empirically supported phylogenetic histories, as well as by the realization that although the former may be helpful heuristically, only the latter count as genuine stories

45

(Kitcher 1985; Rosenberg 1985b). With these caveats, let's weave our theoretically plausible story of the phylogenetic origins of some of our base-level moral capacities.

Only some organisms care for their offspring; humans and primates generally do so. Indeed, many offspring do not need parental care or very much of it. Human offspring and, plausibly, the offspring of our hominid ancestors do. With humans, we know that much of an infant's development comes after birth; the infant cannot survive on its own. What evolutionary reason(s) might account for this? One reason is that, given the increasing size of the hominoid brain, it became increasingly difficult to give birth to an infant whose brain is fully developed. "Premature" birth became a kind of evolutionary necessity; and, along with premature birth came the need to care for newborns and infants. Thus, having the tendency to provide care for young infants who are almost helpless for a number of years after birth would certainly have been advantageous in our hominid ancestors.

Next, it is quite reasonable to believe that as social primates the predecessors of humans and early humans originally lived in small groups of related individuals. Because of this, we might expect selection for traits that benefit not only direct descendants but also indirect descendants. So, besides selection for behavioral tendencies promoting altruistic actions toward direct descendants, we would also expect selection for behavior tendencies inclining their bearers toward helping kin.

So far, so good. But what about tendencies to help nonrelated others? We have already mentioned reciprocal altruism as being the object of neo-Darwinian selection. But how would this work? Why and how would cooperation spring up between nonrelated individuals? Much suggestive theoretical work has been done on this issue, and several scenarios have been laid out. A common starting point has been adopted from game theory, specifically the Prisoner's Dilemma Game. Consider the following situation. Suppose you and I have been imprisoned by the police for stealing. They have us locked up in separate cells, and they're trying to get us to confess. We cannot get word to each other, so we have to make our decisions on our own. Now here's the key point. We're absolutely rational people. "Being rational" is defined by game theorists and economists to mean that we have our own ends and that we choose the right way to get where we want to go. The detectives have informed us that we have the following choices. For me: If I confess and you don't, you get 10 years and I go free. If I confess and you do too, then I get five years and so do you. If I do not confess and you do not either, then I get three months and so do you (because they have no evidence to hold us any longer than that). And if I do not confess and you do, then you get off free and I get 10 years. For you, the situation is exactly the same: If you confess and I don't, then I get the sucker's payoff: I get 10 years in the cooler and you get off scot free. But if you don't confess and I don't either, then we both get the cooperator's payoff: I get three months and you get three months. On the other hand, if you don't confess and I do, then you're the sucker. You get 10 years in jail and I go free. And if we both confess, then we each get the confessor's payoff: five years.

What should I do? What should you do? Remember, we can't talk to each other and we're completely rational, that is, we are completely self-interested and will always pick the best means to attain our ends. It's very simple. Take it from my point of view. What's best for me to do, no matter what you do? If you confess, it's better for me to confess than not to confess. And if you don't confess, it's better for me to confess. In the first case I get 5 years in jail, whereas if I had not confessed, I would have gotten 10. And in the second case, it's better for me to confess than not to confess since I get off free, whereas if I had confessed, I would have gotten three months. Exactly the same situation holds for you. It will always be better for you to confess, no matter what I do. The most rational thing for both of us to do is to confess.

Thus the Prisoner's Dilemma Game involves a reward structure in which the highest payoff goes to those who act selfishly and refuse to cooperate, while their partner cooperates. The second highest payoff is received by individuals who both cooperate. The next highest payoff goes to individuals who have both refused to cooperate and the lowest to the altruist when the other player refused to cooperate and acted selfishly. The reward structure captured by the Prisoner's Dilemma Game is one that arguably occurs often in the lives of savannah-dwelling primates and occurred often in the lives of our hominid ancestors, for instance, in such activities as grooming, food seeking, mating, and avoiding prey. Thus, it seems reasonable to use it to figure out whether and under what conditions altruistic, that is, cooperative behaviors, tendencies, and motivations might evolve. Without going into the mathematical details, it has been determined that a key factor in the evolution of altruism depends on whether there is nonrandom interaction between the participants, that is, some correlation between who plays with whom (Sober 1992). Given the proper reward structure with respect to the costs to the donor and the benefits to the recipient, altruism can evolve when the structure of the interaction allows altruists to have the opportunity to interact with each other more often than with selfish individuals. We can envision various sorts of structures that might bring about this correlation between interactors, for instance, (1) spatial distribution, (2) kin groups, (3) certain behavioral/cognitive strategies in repeated, compulsory interactions,

(4) more sophisticated cognitive and behavioral strategies in optional, non-compulsory interactions, and (5) group selection (Sober 1992; Kitcher 1993).

Supposing that we already have kin groups, how can reciprocal altruism toward nonrelatives originate, take hold, and perdure? To put the last question more technically, biologists ask whether and how altruism might be an evolutionarily stable strategy, that is, a strategy that is such that if all the members of a population employ it, then no mutant using a different strategy could invade the population under the influence of natural selection (Maynard-Smith 1982, p. 10). We might suppose that chance has put nonrelated altruists together and that geographical barriers have kept them separated from selfish individuals. But selfish mutants and the immigration of selfish individuals into the altruist group are not unlikely events. In either case, if the members of the group are interacting on a random basis, eventually selfishness will take over. We can see this intuitively. A selfish individual always does better than an altruist, so a selfish individual will prosper in an altruistic community. As selfish individuals increase in the population, they will run into more of their kind, but overall they will continue to do better than altruists. The latter will be meeting more and more selfish individuals and fewer of their own kind, and thus will be doing worse.

But suppose that although the interaction remains random, the number of interactions between individuals increases and the altruist becomes more sophisticated. She adopts a strategy of cooperating whenever she first encounters someone, but after that the pattern of her behavior becomes more complicated. In the next round of the interaction, if her partner has cooperated on the first round, she cooperates on the second encounter; if her partner fails to cooperate and defects on the first encounter, then she defects on the next encounter. After that, she does exactly what her partner has done on the previous encounter, either cooperating or defecting. For obvious reasons, this strategy is called Tit for Tat (TFT). If TFTers play with selfish persons who always defect, it turns out that TFT is more successful than the always-defect strategy. We can once again see this intuitively. Even though the interactions between the players are at random, that is, even though it is just as likely that a TFTer will play with one of her kind as with someone who always defects, a nonrandom pattern develops in each round of interactions. TFTers settle into a round of cooperative games and defectors into a round of games in which each defects. And after the first round, TFTers playing with those who always defect play their strategy. The mathematics shows that with a given ratio of cost to benefit, the likelihood that TFT will evolve in the population increases either as the number of games in a round increases or as the proportion of the TFTers in the population increases (Sober 1992).[4] Thus, it turns out that if you want to win an Iterated Prisoner's Dilemma Game, a game where you play the Prisoner's Dilemma over and over with the same partners, the best strategy is not the most self-interested strategy but TFT (Alexrod and Hamilton 1981; Alexrod 1984). Moreover, it is reasonable to assume that most of our morally relevant interactions will occur in situations of interaction over a long period of time. That certainly seems likely at least for our hominid ancestors. It should be noted that "Tit for Tat" remains the best strategy even when everyone knows that everyone else is using this strategy. Once it is in place, the strategy is evolutionarily stable against the strategy of always defect (Maynard-Smith 1982, pp. 202–3). But how is it put in place to start with?

Suppose that organisms are genetically built so that they are other-regarding toward family members and play "Tit for Tat" in cases where there is some uncertainty about whether the other person is a relative or not (Rosenberg 1991). This assumption is more likely than the supposition that individuals have a super kin selection device that enables them to recognize precise degrees of relationship and calculate their behaviors accordingly. For we suppose that evolutionary processes usually solve problems with mechanisms that are early in emerging, are not too costly to build, and are successful in a variety of circumstances. The super kin selection device would surely take longer to emerge than the solution using kin selection, plus "Tit for Tat" in borderline situations, because the former is more complicated. The latter is also less costly since it does not require the multiple generations of trial and error that the former probably requires. In the long run, the super kin selection device would be more successful than the "Tit for Tat" strategy, but it seems reasonable to believe that the latter is flexible and successful enough that, along with its advantages with respect to the other two criteria, it turns out to be the preferable strategy. Suppose now that there is a genetic mutation for playing "Tit for Tat" with everyone or at least with all strangers. Interaction with many different selfish strangers would be very costly, since "Tit for Tat" demands cooperation on the first round. However, it is also reasonable to suppose that such interactions would be relatively rare in early human evolutionary history. We can also reasonably suppose that, for organisms that lack complex cognitive and calculational powers, the costs of playing "Tit for Tat" with everyone are less than those of using more complex strategies. It would cost more to store and maintain the information to adapt a more complex strategy than to be a sucker on the first round of a "Tit for Tat" strategy. This would especially be the case if, as we have mentioned, the likelihood of meeting strangers is rather low. So we have a way of seeing how reciprocal altruism could get started in a population.

Kitcher (1993) proposes what might be a more plausible evolutionary scenario in which the Prisoners' Dilemma Game is an optional one in which a player has the opportunity to engage with some participants but not others and the opportunity to opt out altogether. In addition, he assumes that the strategies used by the players are cognitively and behaviorally more sophisticated. The participants are one of the following types: (1) Discriminating Altruists (DAs), who play with any individual that has never previously defected on them and, when they play, always cooperate; (2) Willing Defectors (WDs), who always play and always defect; (3) Solos (SOs), who always opt out; and (4) Selective Defectors (SDs), who play with any individual that has not defected on them and always defect. The reward structure of the optional game is as follows, in order of highest reward: (1) You defect and your partner cooperates; (2) you both cooperate; (3) you opt out; (4) you both defect; and (5) you cooperate and your partner defects; and (1) is equal to or less than the sum of (4) and (5). Kitcher shows that under these assumptions the prospects for the evolution of biological altruism are encouraging. If the benefits of cooperation are sufficiently large relative to those of opting out, and if the number of opportunities to play the optional game is large relative to the population size, then once it gets going, it can be kept in place. Moreover, a pair of altruists can invade both a population of SDs and one of WDs.[5] Finally, approximating human altruism even more closely, Kitcher increases the cognitive and motivational sophistication of his players by postulating a second-level set of motivations that modifies the way the first-level set of rewards is viewed. The first payoff matrix represents the individual's selfish desires in the sense of what the individual would be moved to do if there were no consideration of the effects on the other. The second payoff matrix takes into account an individual's desires about the well-being of the other (or others). Selfish individuals leave the first matrix as is; altruists modify it. Kitcher then goes on to determine the conditions under which golden rule altruism, the valuing of oneself and another equally, might stabilize.

I have fashioned an evolutionary story that shows how nature can select for various sorts of altruistic tendencies and capacities inclining our hominid ancestors and ourselves toward behaviors beneficial to our offspring, relatives, and nonrelated others. Moreover, these are the sorts of capacities that are both substantively and functionally morally relevant because of the links between biological altruism and substantive morality and the role of these capacities in complex moral agency. Thus I conclude that both neo-

50

Evolution and morality

Darwinian theory and the sociobiological extension of neo-Darwinian theory provide theoretical links between evolutionary theory and moral agency. They make plausible theoretical claims both about the existence of evolutionarily based moral tendencies that are candidates for being components for the base level of moral agency postulated by my model and about how we come to acquire these tendencies. But do we actually have such capacities? And what do they look like, that is, what is their nature? It is to those questions that we now turn.

NOTES

1. Natural selection is not the only force that can change gene frequencies in subsequent generations. Other factors are, for instance, immigration, migration, mutations, recombinations, linkage, the accidents of death, and mating.
2. Darwin himself was aware of this problem, and some, like Alexander (1987), argue that he had an intuitive grasp of its solution even if he did not have the technical concepts to give it a precise formulation. Robert Richards (1987) provides a lucid and compelling account of the role that the problem of altruism played in Darwin's development of the theory of natural selection.
3. Sober (1989, 1993, 1994) has provided a helpful way of understanding selfish and altruistic motivational structures in terms of a two-dimensional preference structure. He assumes that people have self-directed and other-directed preferences. We can imagine, then, that there are four basic sorts of choice-results: (1) the self and other gain; (2) the self loses and the other gains; (3) the self gains and the other loses; (4) both the self and the other lose. Preference structures are determined on the bases of which choice result is most valued. Only some behaviors can distinguish the altruist from a selfish person since, for instance, both those with a basically selfish or basically altruistic preference structure, what Sober calls *extreme selfishness* and *extreme altruism*, would prefer choice result (1) over, for example, (4). However, a choice situation in which the options are either (2) or (3) will enable the distinction between altruistic and selfish motivations.
4. The formula w(TFT) is greater than w(ALLD) if and only if $p(n-1)$ is greater than c/b, where w(TFT) is the fitness of the TFT strategy and w(ALLD) is that of the always-defect strategy, p is the proportion of individuals in the population using the TFT strategy, n is the number of times partners play against each other, and c/b is the ratio of costs to benefits.
5. Altruists can establish themselves against defectors, either directly, if mutants are postulated to arise in a population of SDs or WDs, or indirectly by being able to invade SOs, since SDs and WDs will eventually collapse into populations of SOs.

3

Evolution and moral agency
Does evolution endow us with moral capacities?

Do we actually have evolutionarily based moral capacities? Three prominent contemporary biologists, among others, have addressed that question
in some detail and given it significantly different answers. Although all
three have taken broadly interactionist stances, after that they part ways.
George C. Williams (1988a,b, 1989) has adopted an antagonistic interactionist stance, arguing that biology and morality are in mortal combat with
each other. For Williams, the ends of evolutionary processes, whether the
ultimate ones of survival and reproduction or the proximate ones of various
adaptations, are morally tainted and conflict with morality.[1] Richard Alexander (1979, 1985, 1987) puts morality in the service of biology, viewing
evolutionary ends as morally neutral. E. O. Wilson (1978) places biology in
the service of morality, and argues that they provide the bases for morality.[2]
I shall examine both Wilson's proposal and the support that he brings for it
and argue that Wilson's evolutionary story about morality cannot be the full
story. Although the support that he offers for the existence of evolutionarily
based moral capacities is suggestive, much more evidence is needed to
make his plausible suggestion into a well-supported hypothesis. In the
following chapter, I attempt both to expand Wilson's evolutionary story and
to argue for an integrationist approach by shifting our focus to development
and the ontogeny of moral capacities.

3.1 E. O. WILSON: BIOLOGY IN THE SERVICE OF
MORALITY

E. O. Wilson's most comprehensive account of his views on the biological
bases of ethics is found in his *On Human Nature* (1978).[3] We can summarize the line of his argument there as follows: (1) We have some urgent moral
problems that require solutions. (2) Given our biologically based human

52

nature, we have a range of alternative solutions. (3) We have the ability to choose from among these alternatives; that is, we have biologically based cognitive, emotional, and behavioral capacities to help put these choices into action. (4) Finally, we have a set of basic values deriving from our genetic makeup and biological knowledge that tells us what sorts of choices we ought to make. Premises (1) and (2) speak to the issue of substantive moral relevance. Premise (3) speaks to functional relevance (we have the sufficient knowledge and freedom to be moral agents), acquisition (we have in-built and acquired moral capacities), and action (we have the motivations to put these capacities to work to achieve the moral values in question). Premise (4) addresses the question of adequacy. So it is clear that Wilson is proposing a theory that addresses the questions of moral agency.

Moreover, that proposal is clearly a biologically based one. All six possible connections between evolutionary theory and moral agency are implicitly or explicitly asserted to be true. Clearly, for instance, with respect to each of the biologically based behavioral tendencies that Wilson discusses, those connected with aggression, sex, and altruism, biology provides us with factual information about the circumstances and conditions in which these tendencies express themselves and so provides us both with understanding and with valuable knowledge if we want to modify these tendencies or make proper use of them. Moreover, Wilson is offering us an evolutionary explanation of the ultimate origins of our capacities for moral agency and of the proximate motivators of our moral actions, our emotionally funded behavioral capacities.

In some areas, Wilson does not offer us specific normative rules but lays out some of the normative options, for instance, with respect to the relationships between the sexes and types of family unit. On another topic with respect to sex, he is more specific, urging that the morally correct attitude for heterosexuals is to be tolerant of homosexuals. Sex is the topic where Wilson is clearly critical of the "natural law" approach to both the nature and function of sex in humans. The natural law approach has gotten its biology wrong and, as a consequence, is urging normative positions that are unacceptable. Nor can there be any doubt that Wilson maintains that biology has a role to play in metaethics. Wilson proposes three cardinal values that are designed to provide justification for normative decisions and function as ultimate biological values. So morality, in Wilson's view, is not merely instrumental to some nonmoral end but has an intrinsic value. Finally, Wilson attempts to display the meaningfulness of his account of biologically based morality by means of the myth of scientific materialism.[4]

Although Wilson has made proposals with regard to each of the substantive connections between biology and morality, I focus on the explanatory connection, the crucial one for the purpose of answering questions of acquisition and action for moral agency. Our central question, then, is, how adequate is Wilson's claim that we possess evolutionarily based moral capacities? In order to address this question, we must first examine the context in which Wilson presents his claims: the moral dilemmas that we face as a species.

3.2 MORAL DILEMMAS

It is clear both that Wilson realizes the severity of the problems facing the human species and that he believes that these problems are not merely technological, social, and political. They are, in his view, fundamentally moral problems. Wilson is not interested in offering solutions to particular problems such as those of poverty, discrimination, overpopulation, environmental degradation, or species extinction. Rather, he aims to solve the fundamental moral problems that are at the root of the particular problems that press on us from all sides. His contention is that a solution for these problems will be found in a biologically based ethics. In his own words, words that have haunted him ever since, he raises for consideration "the possibility that the time has come for ethics to be removed temporally from the hands of the philosophers and biologicized" (1975, p. 562).

Wilson can be read as posing the question of values, the fundamental basis for our ethical choices, in terms of a series of historical dilemmas facing the human species.[5] The first historical dilemma, which has arisen only in modern times, goes as follows:

Either our values come from God or they do not.

If our values come from God, then values are dictated to us and are extrinsic to us.

And if our values do not come from God, then there is no basis for values.

Thus either our values are dictated to us and are extrinsic to us or there is no basis for values.

Wilson believes that this dilemma has been solved by denying the implication of the second horn of the dilemma. He argues that it does not follow that if there is no God, there is no basis for values and thus for our ethical choices. Our values can be based on human nature.

But this solution only leads to another dilemma because our human nature, in Wilson's view, is not itself a unitary and harmonious whole. Thus

Evolution and moral agency

Wilson believes that having solved the first historical dilemma, we are now faced with the following second dilemma:

> Either values are based on our human nature or they are not.
>
> If values are based on our human nature, then they are logically incoherent, since eventually we will be involved in circular reasoning in the determination of what is valuable.
>
> If values are not based on our human nature (and since, granting the solution to the first dilemma, we know that they are not theistically based), then they have no basis and hence are arbitrary.
>
> Thus, our values are either logically incoherent or arbitrary.

The charge of logical incoherence in the determination of our values is based on the claim that if our values are founded on our human nature, then we must make moral choices among conflicting biologically based tendencies. How are we to choose between these conflicting tendencies? *Ex hypothesi,* we have only biologically based human nature to serve as a basis. But since human nature is itself not a harmonious whole, it can provide no principle of ordering. Thus, our choices of basic values are condemned to incoherence. As we shall see in more detail shortly, Wilson believes that the solution to this dilemma can be found by denying the implication of the first horn of the dilemma, that is, that a biologically based system of values would necessarily involve circular reasoning. He argues that we can discover biologically based *cardinal* values that can both form the basis for the determination of secondary values and provide the premises for noncircular moral reasoning.

But the solution of this second dilemma will still leave a much more difficult dilemma for future generations. This dilemma is rooted in our power to manipulate and change our genetic constitution. The dilemma takes this form:

> In the future, we will either engineer our genetic makeup or we won't.
>
> If we do engineer our genetic makeup, then the decisions we make will have no ethical basis because we will be changing the very criteria in terms of which we have hitherto (given the solution of the second dilemma) made our ethical decisions, that is, human nature as constituted by our present genetic makeup.
>
> If we do not engineer our genetic makeup, then there still is no fundamental basis for our ethical decisions since our present genetic makeup is itself an accident of evolutionary history.
>
> Therefore, in either case, our ethical decisions have no fundamental bases.

Wilson does not attempt to solve this last dilemma, although it is clear that it raises major theoretical questions with regard to a biologically based system of values. For even if Wilson is correct that there is a set of cardinal values founded in human nature, scientifically considered, nevertheless it remains a fact that evolutionary history tells us that human nature is the product of a process that itself seems to have neither necessity nor a line of direction. And thus, when humans attain the technical capacity to intervene in and change that process, as we are rapidly doing, then it seems that all the directives of a value system premised on the stability and survival of the human gene pool are relativized and we are left without a moral rudder.

Although both the first and third dilemmas are worthy of consideration, let's focus on the second, since it is of immediate concern for understanding and assessing Wilson's claims about the existence and nature of biologically based moral capacities.[6]

3.3 EVOLUTIONARILY BASED MORAL CAPACITIES AND THE SOLUTION OF THE SECOND MORAL DILEMMA

According to the second dilemma, we are faced with the uninviting option of choosing between a set of values that has no basis or a set of values whose basis is logically incoherent. As I have mentioned, Wilson attempts to blunt the first horn of this dilemma, the one that aims to pin him with the fallacy of circular reasoning. He grants that human nature is not a harmonious whole and that it is composed of genetically based and emotionally funded motivators that drive us in divergent directions. He also maintains that a value system based on human nature will have to draw from that apparent moral disharmony. Nevertheless, Wilson contends that this can be done in a noncircular fashion. The trick is to uncover the evolutionary source of these divergent motivators and in that way to find the unity in their diversity. That unity is, in Wilson's view, the preservation and continuation of the entire human gene pool.

Thus, Wilson proposes that we can discover three biologically based cardinal values that, he argues, can both form the basis for the determination of supportive secondary values and provide the premises for noncircular moral reasoning. These cardinal values are (1) the survival of human genes in the form of a common gene pool over generations; (2) the maintenance of diversity in the human gene pool; and (3) universal human rights. On the basis of these cardinal values and supportive secondary values, choices can be made between our conflicting genetically based tendencies (1978, pp. 199–200).

How does Wilson arrive at these cardinal values, and what is their biological basis? His discussion of altruism is central to answering this question. Wilson distinguishes between what he calls *hard-core* and *soft-core altruism*. Hard-core altruism refers to an irrational behavioral tendency to aid others, with no desire for an equal return and no attempt, conscious or unconscious, to gain such a return. The origin of hard-core altruism is in kin selection or group selection. It is, in Wilson's view, relatively unaffected by social reward and punishment after childhood. Soft-core altruism, on the other hand, is the tendency to help another, relative or not, with the expectation of reciprocation. Soft-core altruism originated in individual selection but, as opposed to hard-core altruism, is highly affected by culture. Nevertheless, soft-core altruism is also marked by a strong emotional response. This is evidenced, Wilson contends, in the moral aggression expressed in enforcing reciprocation.

Wilson asks us to consider a spectrum of behaviors, beginning with those that benefit the individual and then extending to those that favor the nuclear family, the extended family, the band, the tribe, the chiefdom, and, finally, still larger political units. On this spectrum, human behavior is closer to the side of self-serving action – not as self-serving as, for instance, that of sharks but not at all like that of the colony-centered organisms, like the siphonophore jellyfish or the social insects, for which kin selection predominates. Wilson contends that if hard-core altruism had predominated in humans, the ties of kin and ethnicity would be almost impossible to break; and, in addition, we would see intermediate-sized social unities, like the extended family and the tribe, continue to predominate at the expense of the individual, on the one hand, and nations, on the other. But this is not the case. To show this, Wilson refers to studies that seem to indicate that ethnicity takes second seat to socioeconomic class interest and the latter to individual interest.

The demand for reciprocation that is a part of soft-core altruism is reflected in the absoluteness of moral rules, whereas the protean character of the range of its recipients and alliances is reflected in the varying views about those who are members of the in-group and potential objects of reciprocally altruistic actions. Because of the predominance of soft-core altruism, Wilson believes that humans are capable of indefinitely greater harmony and social stability. Are they capable of "perfect altruism?" Wilson thinks not, although he is not completely clear about what he means. We might take him to mean, by perfect altruism, acting altruistically toward any stranger without expectation of reward. Wilson asks us to consider Lawrence Kohlberg's research on moral development.[7] Whether correctly or not, he reads Kohlberg's stages in such a way that stage five of moral development involves the notion of reciprocal altruism and stage six that of perfect altruism. If Kohlberg's views turn out to be correct, then, Wilson argues, we can think of the first five stages as reflecting the ontogeny of morality in the individual. Individuals are steered by learning rules shaped by natural selection, so that most of them reach stages four and five and are prepared to live harmoniously in a society. However, this society is that of the hunter-gather type of our Pleistocene ancestors; most, if not all, humans no longer live in such a selective environment. Can we move to stage six, the stage of perfect altruism? If we rely on the mechanisms shaped by our earlier hunter-gather environment and built into our genes – and thus our emotional and behavioral tendencies – the answer is no. Wilson reasons that principles chosen on the basis of intuition will be guided by our emotional tendencies and so will reflect their biological bases. As such, they will reflect the mix of hard-core and soft-core altruism that structures our relationships now.

But, then, how is it that we can have, as the first cardinal value of our biologically based morality, the preservation of the common human gene pool? It is hard to see that we can, since it seems that the common human gene pool extends beyond the range of even reciprocal altruism. Obviously, we cannot interact with all members of the species. Nor can we even reasonably expect reciprocation from all those who are indirectly benefited by our altruistic actions. Perhaps cultural factors can move us beyond these biological limits. Wilson indicates that this is at least theoretically possible: "But to the extent that principles are chosen by knowledge and reason remote from biology, they can at least in theory be non-Darwinian" (1978, p. 167). Nevertheless, Wilson concludes that this cannot happen. "This leads us ineluctably back to the second great spiritual dilemma. The philosophical question of interest that it generates is the following: Can the cultural evolution of higher ethical values gain a direction and momentum of its own and completely replace genetic evolution? I think not" (1978, p. 167). Wilson believes that this is so because culture, although not tightly bound by the genes, nevertheless is bound by them.

> The genes hold culture on a leash. The leash is very long, but inevitably values will be constrained in accordance with their effects on the human gene pool. The brain is a product of evolution. Human behavior – like the deepest capacities for emotional response which drive and guide it – is the circuitous technique by which human genetic material has been and will be kept intact. Morality has no other demonstrable ultimate function. (1978, p. 167)

But what, then, permits us to formulate as a cardinal value the survival of human genes in the form of a common pool over generations? Wilson asks us to consider "the true consequences of the dissolving action of sexual reproduction and the corresponding unimportance of line of descent" (1978, p. 197). If we do this, we will realize that our genetic inheritance quickly disperses into hundreds of ancestors as we move back into the past a few hundred years and even into millions of ancestors in about a thousand years. A similar dispersal happens when we consider the future. Thus the human individual, each of us, is "an evanescent combination of genes drawn from this pool, one whose hereditary material will soon be dissolved back into it" (1978, p. 197). This vision, provided by biology, should allow us to take as our primary value the preservation of the human species. It does not seem plausible to read Wilson as saying that only some humans should be preserved. Do the mechanisms of hard-core and soft-core altruism allow for this extension to the entire human gene pool? It seems not. "Because natural selection has acted on the behavior of individuals who benefit themselves and their immediate relatives, human nature bends us to the imperatives of selfishness and tribalism" (1978, p. 197). However, *biological knowledge* allows us to extend the range of what is valuable. "But a more detached view of the long-range course of evolution should allow us to see beyond the blind decision-making process of natural selection and to envision the history and future of our own genes against the background of the entire human species. A word already in use intuitively defines this view: nobility" (1978, p. 197). How, then, do the genes keep culture on a leash? What remains primarily valuable when we move from in-built assessors of value, hard and soft-core altruism, to culturally based ones of biological knowledge, is the entire human genetic endowment, not just some individual human genotype or some particular genetic line. Biological knowledge expands the circle of altruism. Moreover, it seems that people can use their soft-core altruistic tendencies to extend their ties in such a way that more and more people are included as potential recipients of altruistic behavior. And biological knowledge can help us to understand that their reciprocation need not be to us, our children, or relatives in order to be a genuine reciprocation. Thus, cultural forces work with biological forces both to identify what is valuable and to provide the mechanisms by which they are achieved.

We can, I think, understand the second and third cardinal values as instrumental to the preservation of the common human gene pool and also as enunciating values that allow for human excellence, that is, not only for its survival but also for its well-being. The secondary values, in Wilson's view, are those that serve to promote the primary values. Those mentioned by Wilson are ones that he has picked out as having "historically served as the enabling mechanisms for survival and reproductive success" (1978, p. 199).

These values are defined to a large extent by our most intense emotions: enthusiasm and a sharpening of the senses from exploration; exaltation from discovery; triumph in battle and competitive sports, the restful satisfaction from an altruistic act well and truly placed, the stirring of ethnic and national pride; the strength from family ties; and the secure biophilic pleasure from the nearness of animals and growing plants. (1978, p. 197)

Since this list is a historically conditioned one, I take it to be Wilson's view that what counts as secondary values will depend on the particular situations in which the primary values are to be achieved.

So Wilson's solution to the second dilemma seems to go like this. Consider a range of pressing problems like national and ethnic conflict, racial and sexual discrimination, and unjust socioeconomic differences. These threaten us individually, as well as threatening our children, our relatives, and the groups of which we are members. They clearly are problems that have an ethical component, since they have to do with the distribution of benefits and harms and, thus, with how we are to treat each other. Our evolutionary history tells us that we have a number of genetically based behavioral tendencies that find expression in our emotions. These incline us in ways that have, in the past, promoted our evolutionary success, our survival and reproduction. But how, in the present circumstances, shall we choose between these tendencies, since they lead us in different directions? Some lead us to promote individually selfish behavior, others to favor kin or reciprocally altruistic behavior. Others lead us to act selfishly or aggressively toward members of the other sex or other ethnic, socioeconomic, or national groups. If we have no transcendent values system and we want to avoid arbitrary choices, then it seems that we must rely on our human nature. But that's the problem and the source of the dilemma, since our nature seems to be a hodgepodge of emotional and behavioral tendencies. If we choose on the basis of one particular facet of our nature, using the emotional motivators that support it, we seem to be caught in a vicious circle. For instance, we attempt to address the question of socioeconomic differences by saying that there ought to be a more equal distribution of goods because that will increase the benefits for ourselves and for our direct and indirect descendants. When we are asked why that latter goal is something to be pursued, our only reply is that it leads to a more equal distribution of goods. But, Wilson suggests, if we adopt a set of cardinal values, then we can make a decision to bring about a fairer distribution of goods by appealing to the first and third cardinal values. The first demands that we consider all humans and the third that we give them equal consideration insofar as it is relevant to the goods being distributed. So we can avoid a logically circular reasoning and, in principle, resolve the second dilemma.

Thus, reason, based on sound biological knowledge and supported by emotion, itself founded in our biologically based tendency toward soft-core altruism, can, in Wilson's view, provide the basis for moral agency. Biology serves morality. But does it, and if so, how and how well?

3.4 ASSESSING WILSON'S PROPOSAL FOR EVOLUTIONARILY BASED MORAL CAPACITIES

A critic of Wilson's proposal for evolutionarily based moral capacities can question whether the capacities invoked by Wilson are either moral or evolutionary. Let's examine each of these questions. First, are the capacities that Wilson invokes moral? As we have seen, for something to be moral, it must meet both substantive and functional criteria of morality. We have already indicated that a part of the ordinary functional conception of moral agency is that a moral agent must be a free agent. Wilson has two views on freedom, an official one and an implicit, perhaps unrecognized, one (Rottschaefer 1983b). His official one is that individual freedom is a matter of unpredictable events, perhaps of the fundamental ontological sort that one garners from a realist interpretation of quantum indeterminacy. Since, on this view, free choice lacks the understanding, knowledge, and intention requisite for moral responsibility, it appears inadequate for meeting the requirements of moral relevance. On the other hand, Wilson's implicit view, which runs as an integrating theme throughout *On Human Nature,* is that we must make choices about the use of our evolutionarily based moral capacities on the basis of biologically informed knowledge about the nature and consequences of acting on these capacities and on the basis of our biologically informed grasp of cardinal values. The implicit view, therefore, seems to be close to the ordinary view of freedom and, thus, seems to fulfill the criterion of functional moral relevance. That leaves us with the question of substantive moral relevance.

To answer this question, we have to ask what the ends of evolutionarily based capacities are. Consider evolutionarily based moral capacities for biologically *selfish* actions, that is, actions whose ultimate aim is individual survival and reproduction. These actions might be getting enough food to eat, enough rest, proper shelter, defending against predators, finding a mate, caring for offspring, and so on. We have noted that none of these need be selfish in the ordinary sense of aiming to benefit only the individual. Biologically speaking, they are not aimed merely at benefiting the genotype. Rather, insofar as the evolutionarily based moral capacities are adaptive, they are such because they benefit an individual's direct descendants either directly, as in parental care, or indirectly, by benefiting the parent. Are such actions substantively within the moral realm? If we adopt a non-question-begging but relatively minimal criterion for what is within the moral realm, namely, avoiding harming others and helping others, these evolutionarily based capacities are within the realm of morality. Using this criterion, both hard-core and soft-core altruism are also within the moral realm. Since the three cardinal values are all extensions of the ends of altruistic evolutionarily based moral capacities, it is clear that they also fall within the realm of the moral. But even independently of their connection with altruistic, evolutionarily based capacities, their content clearly falls within the realm of morality. The preservation of the human species in its diversity and individuality is commonly held to be a moral good. Given that some of the capacities that Wilson is postulating are substantively moral capacities and that their functional relevance can be established, I conclude that at least these capacities are indeed moral capacities. But now the question arises about their evolutionary character. We can grant that we have moral capacities without granting that they are either entirely or in part *evolutionarily based.*

On what bases would we have reason to believe that these moral capacities have, at least in part, an evolutionary basis? Recall that in our account of evolutionary theory as a selection theory, we saw that the interaction of organisms possessing or not possessing a certain property (or possessing it in varying degrees) with a critical factor (or factors) in the environment of those organisms leads to a benefit for some of those organisms and less or no benefit to others. This, in turn, leads to the increased survival and reproduction of the benefited organisms and, in the longer term, to an increased presence in the gene pool of those genes responsible for the crucial property. In our case, the crucial postulated property is a moral capacity. The critical environmental factors are those factors in the evolutionary history of humans that selected for these moral capacities. What has to be shown is that the environment has selected for organisms possessing moral capacities in preference to organisms with different sorts of capacities, and that that has led to the relative reproductive success of these organisms. To be more specific, consider the evolutionarily based moral capacities we have been focusing on, the capacities for self-care and parental care leading to reproductive success, so-called biologically selfish behavior, as well as capacities for kin altruistic behavior and for reciprocally altruistic behavior. Notice, too, that when we are concerned with something like an evolutionarily based moral capacity, we are interested in a generally shared capacity of humans, not in genetically based individual differences. Of course, there will be individual variations in a population, so we will expect to see that the property displays a normal distribution. So, evolutionarily based moral capacities, if they exist, are not based on genes that are relatively rare in a population – for instance, genes that predispose to some relatively infrequently occurring physical disease or psychological disorder.

Let's use the following sketch of an argument to consider how Wilson attempts to establish that we have evolutionarily based moral capacities and thus an evolutionarily based moral agency.[8] The first three steps follow the general logic of an attempt to establish an evolutionarily based phenotypic characteristic. The last incorporates a way to understand biological human nature by means of the concept of the norm of reaction.

I. If humans possess fitness-enhancing moral capacities in a given critical environment, then we would expect both (biologically) selfish fitness-enhancing behaviors (promoting, for instance, parental care) and (biologically) altruistic fitness-enhancing behaviors toward kin and non-relatives (when reciprocation in that environment from the nonrelatives can be anticipated to be fitness enhancing).

There is evidence for such fitness-enhancing behaviors.

So, there is some reason to believe that humans possess such fitness-enhancing moral capacities.

II. If natural selection is operating to select for moral capacities, then there should be evidence that there are fitness-enhancing moral capacities.

There is some evidence that there are fitness-enhancing moral capacities (from Step I).

So, there is some reason to believe that natural selection is operating to select for fitness-enhancing moral capacities.

III. If natural selection is operating to select for fitness-enhancing moral capacities, then there are genes responsible for these capacities.

There is some evidence that natural selection is operating to select for fitness-enhancing moral capacities (from Step II).

So, there is some reason to believe that there are genes that are responsible for fitness-enhancing moral capacities, that is, that there are evolutionarily based moral capacities.

IV. If humans have evolutionarily based moral capacities, then they have an evolutionarily based moral nature and moral agency.

There is some evidence for evolutionarily based moral capacities (from Step III).

So there is some reason to believe that humans have an evolutionarily based moral nature and moral agency.

This four-step argument, if successful, allows us to conclude that we have evolutionarily based moral capacities and that our moral nature and moral agency have an evolutionary basis. However, as it stands, this argument is still too general to do the job. If Wilson wishes to reach his destination successfully, he has to be much more specific about the crucial elements involved in both the explanans and the explanadum, for instance, the biologically based selfish and altruistic behaviors, the organisms that display them, the proximate mechanisms by which they are effected, and the critical factors in the selecting environment.

Wilson does discuss what seem to be somewhat more specific evolutionarily based morally relevant capacities, for instance, aggressive tendencies, xenophobic inclinations, tendencies relating to the way the sexes relate, tendencies governing the nature and structure of the family and sexual preferences, and tendencies with respect to incest avoidance. We should note that not all evolutionarily based moral capacities need to be considered intrinsically morally good or morally good in all circumstances. Indeed, that is part of the second dilemma that Wilson believes that we are confronted by and that he attempts to resolve. The altruistic behaviors, in the ordinary sense of the term, that is, parental care (biologically selfish), kin altruism, and reciprocal altruism, which exemplify the evolutionarily based moral capacities in our schematic argument, may well be place-holders for much more specific behaviors and behavioral tendencies. On the other hand, they may take the place of still more generalized tendencies. For instance, it has been suggested that instead of discussing a specific sibling incest avoidance tendency or even an incest avoidance tendency, we should think in terms of in- and out-group associative and mating tendencies (Bateson 1989; Kitcher 1990).

Wilson does not fill in all the details of our argument schema by a long shot. Nor does he attempt to delve deeply into the evidential bases needed for the conclusion of our schema. What he offers is a series of hypotheses to establish some of the premises in the preceding argument schema. Although he does not explicitly lay out these hypotheses for us, I find him appealing to the following hypotheses in the course of his attempts to establish an evolutionary account of our human nature:

I. *Interspecies Comparison Hypothesis:* (1) If some human social behaviors are genetically based, then we would expect to find these social behaviors shared by our closest noncultural primate relatives. (2) If all human social behaviors are culturally based, then we would not expect to find similar social behaviors in primate relatives that do not have culture (1978, p. 20 et alibi).

II. *Cross-Cultural Comparison Hypothesis:* If some human social behaviors are genetically based and not just culturally based, then we would expect to find some common social behaviors in all human cultures (1978, p. 20 et alibi).

III. *Social Experiment Hypothesis:* If there are genetic limits to certain social behaviors, then attempts by societies to break or exceed these limits should ultimately fail (1978, pp. 36–9; pp. 80–1 et alibi).

IV. *Genetic Basis Hypothesis:* If some human social behaviors are genetically based, then as technology becomes available, we should expect to find the genes that determine these behaviors (1978, pp. 43–4 et alibi).

Hypotheses I, II, and III are aimed at establishing generally shared phenotypic characteristics; they are hypotheses that can be used to discern the presence of behaviors that may be rooted in evolutionarily based moral capacities. Hypothesis IV represents the ideal culmination of the search for evolutionarily based capacities. One would expect that if there are genes for certain capacities for social behaviors, in our case capacities for moral behaviors, then we would eventually be able to trace the complicated pathways from genes to capacities.

Consider Step I of Wilson's argument. This apparently simple step turns out to be complex. The biologically selfish behavior that we are interested in is one that is directed toward survival and reproduction. How do we determine which behaviors actually serve the function of survival in the service of reproduction? For instance, do all of our efforts to build up and maintain our health count? What sorts of self-enhancement increase fitness? What are the measures of relative fitness enhancement? How do we account for the fact that not everyone mates and has children? That not everyone engages in parental care? That there is child neglect and child abuse? What sorts of parental care best enhance one's fitness? Are there particular evolutionarily based moral capacities for different ways of caring for children, for instance, being warm and attentive or being cold and distant? Moreover, we have to remember that more than likely these capacities were developed a million or so years ago in environments probably very different from modern ones. The behaviors resulting from these capacities may have been fitness enhancing then, in those environments, but not now. Or, as we shall see when we come to Step II, behaviors that are currently fitness enhancing need not be the result of evolutionarily based capacities.

Wilson does not address these sorts of difficulties. That is not to say that they cannot be answered. Consider a kind of behavior that from an ordinary perspective would be considered to be substantively moral, like parental care. From an evolutionary perspective, it is not necessary that everyone be a parent and care for his or her children. We would expect a normal distribution with regard both to being a parent and to involvement in parental care. We could explain why some people are not parents or do not care for their children, or do not do so adequately, in a number of different ways. The explanations for not caring for one's children, for instance, could include biological ones, defective genes or genetic mutations. More likely, we could find explanations in terms of the vagaries of individual learning histories. Nor would it be impossible for Wilson to spell out the sorts of benefits that might be associated with fitness enhancement. So, even though Step I is not as simple as it first appears, there is some reason to concede to Wilson, on the basis of both the animal comparison and cross-cultural hypotheses, that there is some support for the second premise of Step I: There is some evidence that biologically selfish and altruistic behaviors are fitness enhancing. But it is clear that Wilson has not made his case. At best he has suggested how a case might be made.

If Step I appeared simple but in fact was complex, Step II is equally complex and is, in addition, perilous. The peril is, as we have intimated, that the evolutionary force of natural selection is not the only source of fitness-enhancing capacities and behaviors. Thus, even though, as premise one of Step II indicates, the action of natural selection, in the absence of other factors such as balancing selection, is a sufficient condition for the presence of fitness-enhancing capacities, it is not a necessary condition. Learning can mimic natural selection. That is to say, someone can, by practice, imitation, or instruction, acquire capacities that are fitness enhancing. Thus, even though we discover that people have fitness-enhancing moral capacities, we are not assured that these capacities are the result of natural selection. Since learning usually takes place in a social setting and a particular culture, let's call the fact that learning can mimic the results of natural selection *cultural mimicry*. The possibility of cultural mimicry alerts us to the fact that evidence for fitness-enhancing moral capacities, the conclusion of Step I and what we use in premise two of Step II, is not sufficient to allow us to conclude that natural selection is operative in the origin of these capacities.

Wilson is, of course, aware of this problem and uses his first three sociobiological hypotheses to show that natural selection is at work. Thus, the first hypothesis attempts to establish that natural selection is at work by providing evidence that the phenotypic characteristic under discussion is shared by closely related animal species. At least one further assumption is necessary in order to use this evidence to conclude that it is likely that the similar characteristic in humans is also the result of the action of natural selection. We must assume that the phenotypic character of the related animal species is itself likely to have arisen by natural selection rather than by culture.[9] Of course, there are numerous examples of parental altruism, kin altruism, and reciprocal altruism in nonhuman animals. Wilson does not try to make the case that these are proto-moral behaviors. In many if not all cases, we can assume that they are not functionally moral. But, given our account of substantive morality, they are substantively moral. However, Wilson has not attempted to show that these shared characteristics are, in some or all nonhuman species, the result of natural selection rather than culture. Nor has he attempted to show that it is reasonable to assume that since the moral capacities are in some way shared, it is likely that those possessed by humans also arise from natural selection.

Indeed, natural selection and cultural selection are not the only ways in which moral capacities could have arisen. Evolutionary theorists tell us that phenotypic characteristics can also emerge through, for example, random genetic drift (i.e., when changes in allele frequency occur because the genes of parents are not a representative sample of those in the population), mutation, migration, immigration, or pleiotropy (i.e., when the allele affects several different characters). But even if it is not likely that an evolutionarily based moral capacity has arisen in one of these fashions, there are two other ways in which evolutionarily based moral capacities may have come about other than through selection. The first is connected with linkage, that is, the phenomenon of the association of genes on a chromosome. One of the alleles of one of these genes may be selected for, and an associated gene, although adaptively neutral or even adaptively harmful, comes along for the ride. It is a free rider. So it could be that evolutionarily based moral capacities are free riders. There has been selection *for* something else but selection *of* evolutionarily based moral capacities, in addition. Second, we have to remember that things can be used for many different purposes. You can buy Wilson's large and magnificent tome, *Sociobiology,* to help you get a better understanding of sociobiology but also use it to prop up all the other books on your shelf. So an evolutionary capacity may have been selected for one function that it performs in enhancing fitness but can now be used by us for another purpose. We might suppose intelligence has been selected for since its capacity to anticipate the future seems to give us an adaptive advantage. But once selected for to serve that purpose, it is now used for a different purpose: to develop a morality (Singer 1981). Thus, it is clear that even if Wilson can use the conclusion of the first step in his argument, namely, that we possess fitness-enhancing capacities, it does not follow automatically that these capacities have arisen from natural selection. Nevertheless, as with the first step of his argument, so, too, with this second step, there is some plausibility to the claim that our behavioral, emotive, and cognitive tendencies with respect to parental care, kin altruism, and reciprocal altruism were selected for. However, Wilson does not do anything notable to make the empirical case for this.

Step III is relatively noncontroversial since it is a direct consequence of the theory that natural selection works on genetic variation. Of course, as we have seen, other evolutionary forces, such as migration, immigration, and linkage, also change allelic frequencies. So natural selection is not a necessary condition for evolutionary change. Wilson, of course, does not dispute that.[10] We should spend a little more time on Step IV, however, because it concerns the important issue of human nature and genetic limitations. There are two important issues that we should examine in connection with claims about an evolutionarily based moral nature and moral agency. The first has to do with what we might mean by human nature as such, and the second is concerned with the determinants of human nature and, in particular, its evolutionary determinants.

How can we understand human nature? If we assume that humans are evolved animals, we cannot make the classical philosophical assumption that human nature can be identified by means of essential characteristics. Significant human characteristics have links with those of their evolutionary ancestors. Thus, significant human characteristics, including human morality and moral agency, are continuous and not dichotomous. Nevertheless, it remains reasonable to attribute some characteristics to humans, and not to other animals, insofar as humans may possess refinements of capacities that other animals do not. We might, for instance, argue that in the case of agency, even though many animals, especially primates, have a second-level behavioral system, they do not have a third-level reflective system. If that is so, we can attribute agency to both nonhuman primates, for instance, and to humans but in different degrees.

The other aspect of the issue of human nature, that of limitations, is more significant since it is frequently associated with determinism and the restriction of possibilities. In the case of an evolutionarily based human nature, much discussion has focused on the issue of genetic determinism. Wilson often invokes the metaphor of genes holding culture on a leash, albeit a long one. But it is incorrect to think that evolutionary theory somehow implies genetic determinism. If the theory of evolution by natural selection is correct, then both genes and environment are involved in the production of adaptations. Moreover, there is no reason to doubt that natural selection plays a major role in evolutionary change. However, the alternative to genetic determinism is not environmental determinism. As we have seen in our examination of natural selection as a type of selection theory, both genes and environmental factors play crucial roles in determining adaptations. So, if there are evolutionarily based moral capacities and, thus, an evolutionary basis to our moral nature and moral agency, then both genes and environment are factors in their determination. Obviously, then, a crucial question is the range of their relative determination. How genotypically or environmentally flexible is our moral nature and moral agency? The former question, of course, is particularly difficult to answer, since we do not have a very good idea of most of what constitutes the human genome, let alone the genetic basis, if there is one, of our moral capacities. Nor do we have much of an idea of what sorts of genetic changes are possible while maintaining the species.

We can specify more clearly what we mean by an evolutionarily based moral nature and moral agency by using the biological concept of norm of reaction (Futuyma 1979; Kitcher 1985). The norm of reaction will help us understand the range of human potential given the human genotype. We can think of a genotype as developing in a range of environments all of which are such that the organism living in it can survive and reproduce. The phenotypic characters that the organism will display will vary, and for each phenotypic character there will be a range of expression. A typical example is the height of a plant in different types of soil and under various climatic conditions. The range of expression of a phenotypic character of a particular genotype is its norm of reaction. For the different genotypes of organisms of the same type or species we would get a somewhat different norm of reaction for a particular phenotypic character. Suppose that we knew the relevant environmental factors influencing the development of a particular phenotypic character. We could then take each organism, with its somewhat different genotype, and determine the effect of a particular relevant type of environment on the expression of that phenotype. We could then average the effects to get the average effect of that particular environment on a particular kind of organism. Then we could do the same with each of the relevantly different environments in which the organism can survive and reproduce. Thus, we would get the norm of reaction for that type of organism with respect to a phenotypic characteristic. Of course, this is all idealized science, hardly practical even for simple organisms, let alone complex ones. In addition, given generally accepted moral constraints, attempting to establish experimentally a norm of reaction for a human would be ethically prohibited. Nevertheless, we can understand what the norm of reaction portrays, how it displays the interactions between genes and environment that produce adaptive traits. From the norm of reaction we can discern the range and variability in a phenotypic trait.

Indeed, if the phenotypic variation of a character for different genotypes in a similar environment were nonexistent or minimal, we would argue that the differential effect of the genotype on the phenotype was minimal relative to the environmental effect. On the other hand, if the phenotypic variation were large in a similar environment, given different genotypes, we would argue that the differential effect of the relevant environment on the phenotype was minimal relative to the effect of the different genotypes. In the first case, environmental influences make a bigger difference and, in the second, genetic influences.[11]

If human moral agency is based on an evolutionarily based moral nature founded in evolutionarily based moral capacities, then we would expect to find a norm of reaction for human evolutionarily based moral capacities. For instance, if parental care is such an evolutionarily based moral capacity, then we would expect human genotypes in various relevantly similar environmental situations to show phenotypic variation in that characteristic. Moreover, we would expect that at the environmental extremes, a characteristic would also take on its extreme forms, perhaps minimal and maximal cognitive, affective, and behavioral tendencies with respect to child care. In addition, we would expect that, in comparison with other related species, there would be some overlap of the curves representing the norm of reaction of the different species. Thus, norms of reaction can display the flexibility and the limits of a phenotypic character. Theoretically, they could do so with respect to our evolutionarily based moral capacities. Thus, we can think of human nature, including our moral nature, as a composite of phenotypic characteristics. In the case of our moral nature and moral agency, these characteristics would consist of cognitive, affective, and behavioral tendencies. These tendencies are both genetically and environmentally based. A more refined analysis distinguishes various sorts of environmental influences. Some of these are relatively constant, for instance, those often associated with ontogeny. These, then, we assign to the developmental environment. Others are more variable but still relatively stable because they invoke learning mechanisms that are pervasive and that are less cognitively penetrable than other, cognitive learning mechanisms. The former belong to the "noncognitive" learning environment and the latter to the cognitive one. Other environmental factors are still more various; these we might associate with social or cultural environments. What this still rough-and-ready refinement of environments indicates is that we need to conceive of our moral agency and moral nature as biological, developmental, psychological, social, and cultural. It also points to a way of examining the acquisition and implementation of our moral agency that we shall be following in subsequent chapters.

70

Evolution and moral agency

We have now examined in some detail Wilson's argument for the existence of evolutionarily based moral capacities. Although the theoretical
assumptions that we have had to make are not implausible, at this point we
do not have much more than a plausible story. It may well be the case that
we have been endowed by evolution with affectively and cognitively motivated behavioral capacities relevant to the moral domain.[12] And it may be
that these form part of what we have called the base level of moral agency.
But we need more evidence than that offered by Wilson. Even if we grant
that there is an evolutionary story to tell about why and how we have
acquired our moral capacities, that is, why and how we have evolutionarily
based moral capacities, there is more to say about how these evolutionarily
based moral capacities are put into *action*. What role do they play in the
complex four-level system that we have hypothesized to constitute moral
agency? How do they motivate and why? These questions are what biologists and evolutionary theorists call *proximate* questions. If evolutionary
theory has given us a sketch of some answers to ultimate questions about
why and how, and so some answers to ultimate questions about acquisition
and action, we still have proximate questions about how and why evolutionarily based moral capacities are acquired and put to work in the service
of moral agency. Wilson's account has not moved us any further with the
question of action.[13]

To make the case for an evolutionarily based moral nature and moral
agency, we need to know more about our moral capacities. We need to
understand more clearly the nature of these capacities and their evolutionary
basis. To do that, I turn to developmental biology and psychology to see
what contributions these disciplines can make to our understanding of the
ontogeny of these capacities. I hope that by obtaining a better knowledge of
how an individual acquires her moral capacities, we will achieve a better
understanding of both their nature and their evolutionary character.

NOTES

1. Williams's account of morality is problematic both substantively and functionally. To make his case that nature is fundamentally at war with morality,
   Williams assumes that reproduction and survival, as well as the adaptations
   promoting them, are fundamentally immoral because they are selfish. I believe
   that one can reject Williams's views about the immorality of nature without
   becoming hopelessly romantic about it.
2. In order to make his case, Alexander must assume that the ends and mechanisms
   of morality are generally subordinated to those of evolution and that the ends of
   the latter are amoral. But he makes no effort to show that reproduction and
   survival, or the adaptations that serve them, are substantively amoral; in fact, he

denies that biology can say anything about this issue. On the other hand, his case for the general subordination of the mechanisms of morality, as well as psychological and cultural factors generally, to biological forces is extremely problematic (Kitcher 1985; Bradie 1994).

3. In the preface to his book *On Human Nature* (1978), Wilson tells us that although he did not plan it, *On Human Nature,* is in fact the final volume of a trilogy of interconnected views on the social traits of organisms. In the first of these volumes, *The Insect Societies* (1971), Wilson analyzes in detail the social systems of the social insects. That volume concludes with a chapter on "The Prospect for a Unified Sociobiology," in which he surveys the prospects for applying the principles of sociobiology deriving from population genetics, ethology, evolutionary theory, and other biological disciplines to vertebrate animals. Those prospects were fleshed out in ample detail in his *Sociobiology: The New Synthesis* (1975), in which he brings together a vast amount of the sociobiological research on the social behaviors of vertebrates. That book, too, ends with a forward-looking final chapter, "Man: From Sociobiology to Sociology," which anticipates the next step in Wilson's intellectual odyssey. In that chapter he speculates about the possibilities for a human sociobiology, a discipline that would link the biological and social sciences. *On Human Nature* is the attempt to spell out in speculative fashion such a human sociobiology. Although the conclusion of *On Human Nature* anticipated no further development of his line of thought, in fact, the debate provoked by his work and that of other sociobiologists led Wilson to write two more volumes. In 1981, in cooperation with Charles Lumsden, he presented in *Genes, Mind, and Culture* an evolutionary theory of mind and a coevolutionary theory of genes and culture to fill in the gap between genes and social behaviors that his discussions in *On Human Nature* seemed to have left glaringly unfilled. And two years later, he and Lumsden published *Promethean Fire,* a popular version of some of the themes of *Genes, Mind, and Culture,* linking them to the ethical concerns that are central to *On Human Nature.* Lumsden (1989, for instance) has continued to fill in the gaps between genes and behaviors with an evolutionary theory of mind and culture. Since it is in *On Human Nature* that Wilson most systematically develops his views on a biologically based morality, I focus my attention there.

4. His idea is to make use of the motivational resources of religion while changing its cognitive content. Since his view of morality is humanistically rather than transcendentally based, he chooses the symbol of Prometheus. Prometheus represents the quest for human excellence and offers hope for a successful quest, although, as we shall see, the nature of the ultimate goals, if any, of that quest remains unclear in the light of the third of his moral dilemmas, the one that Wilson leaves to future generations for solution.

5. Wilson speaks of dilemmas, but his formulations do not take the logical form of a dilemma. I have reformulated them into that form without, I believe, distorting his claims.

6. Wilson's third dilemma surely poses major theoretical moral problems that are

rapidly becoming urgent practical problems. I have examined Wilson's solution to his first dilemma elsewhere (Rottschaefer 1985).

7. We return to Kohlberg in Chapter 6.

8. I have adapted this sketch from Kitcher's analysis of Wilson, what Kitcher calls the *revised standard version* of Wilson's ladder (Kitcher 1985, pp. 122–3, 1986).

9. There are some complications in applying the animal comparison hypothesis concerning generalizations that include *Homo sapiens* and other animals. On the one hand, one could argue that humans have behavioral tendencies similar to those of primates, especially great apes, because we share with primates a common ancestor that passed these traits down to both *Homo sapiens* and primates. On the other hand, one could argue that humans have similarities to some social carnivores because similar evolutionary forces have independently shaped the behavioral tendencies of each lineage even though the lineages are distinct. The commonality of the phenotypic character is grounded differently in the two cases: The first is a commonality due to common descent, and the second is a commonality due to similarity of ecological forces shaping distinct lines of descent. The problem comes in trying to find consistent generalizations about commonalities. For it may well be the case that when we compare humans with primate relatives, we find that they share behavioral tendencies of one type, for instance, nonaggressiveness, whereas if we compare them with an ecologically similar species, we find that they share the behavioral tendency of aggressiveness. Wilson, unfortunately, does not always attend to the need for consistent generalizations about commonalities. See Kitcher (1985, pp. 158ff.) for further discussion of this issue.

10. Despite the critics' charges, it seems clear to me that Wilson is not guilty of the charge of adaptationism, that is, the claim that all traits of an organism are fitness enhancing and have been selected for.

11. There are also the more complicated situations in which there are interaction effects.

12. I have focused on the capacity for parental care as an example of an evolutionarily based moral capacity. For further discussion of this capacity and other possible evolutionarily based moral capacities, see Barkow, Cosmides, and Tooby (1992).

13. Wilson is not completely uninterested in this question; indeed, his discussion of the emotional motivators, originally tied to religious myths and how they might be harnessed afresh by the myth of scientific materialism, is concerned with these sorts of proximate questions.

4

Developmentally based moral capacities
How does the moral sense develop?

4.1 WHAT DOES AN EVOLUTIONARILY BASED CAPACITY LOOK LIKE?

Granting the scientific plausibility of the claim that evolutionarily based moral capacities exist, what can we learn about them and their acquisition by studying their development? Consider two approaches to understanding the nature of these moral capacities as adaptive proximate mechanisms, one straightforward and the other indirect. On the straightforward approach, the identification of different fitness-enhancing patterns of moral behavior, for instance, patterns of parental care for offspring, kin altruism, and reciprocal or group altruism, leads to an inference to a disposition for such behaviors, namely – to use the preceding examples – dispositions to care for offspring, kin altruism, reciprocal altruism, and group altruism. These four dispositions would constitute the totality of our moral agency or a significant part of it and, as evolutionarily based moral capacities, would be the adaptations for which there has been selection. But, of course, things cannot be that simple; even Wilson and Alexander implicitly recognize that. Any one of these simple dispositions, at least in higher organisms and certainly in humans, either must be associated with other proximate mechanisms that are cognitive and motivational in nature or must themselves be complex dispositions with cognitive and motivational components. A bare dispositional account of the nature of the proximate mechanisms that constitute our evolutionarily based moral capacities, if there are such, is unsatisfactory. Indeed, we have seen that the use of game theory to spell out phylogenetic scenarios for the evolution of altruism involves the postulation of rather sophisticated strategies demanding comparatively complex cognitive and motivational capacities. Since a realistic, as opposed to an instrumentalist, understanding of these strategies is arguably the preferable one, a bare dispositional account of the nature of these proximate mechanisms is insuf-

74

ficient. Thus, taken by itself, the simple approach, which begins with differential fitness analyses of patterns of behaviors and postulates bare dispositions for each of the these patterns, is not satisfactory for understanding the nature of our evolutionarily based moral capacities.

On the other hand, rather than starting with fitness analyses and making use of the theories of evolution and natural selection to provide theoretical accounts of ultimate ends and the causal factors operative in the selection of proximate mechanism and ends, one can start with the complexities of our moral behaviors. Starting there, we may be led to postulate a set of complex moral capacities. Here the task is, given the evidence from cognitive and behavioral psychology about our moral learning, and from both developmental biology and psychology about our moral ontogeny, to disentangle the levels of acquisition and their results in the hope of discerning those general sorts of capacities, including moral capacities, if there be such, on which and with which are built our higher-level moral capacities and what we have called our moral agency. Kitcher (1990) has called this alternative, indirect approach *developmental decomposition.*

Direct and indirect approaches are both necessary and complementary. In general, it seems fair to say that our biologists of morality have not made sufficient use of the method of developmental decomposition to establish the existence of evolutionarily based moral capacities and to understand their nature. This method requires an examination of the relevant results from cognitive and behavioral psychology and from developmental psychology and biology. Attention to such results, it is hoped, will not only provide more evidence for the existence of evolutionarily based moral capacities but also furnish much needed clarification of their nature, as well as their modes of acquisition and the ways in which they are put into action.

Patrick Bateson (1989), a respected British developmental biologist, offers us a nice example of this sort of investigation and one that is of particular interest to sociobiologists, especially those who have integrationist inclinations when it comes to the relationships between biology and moral agency. Bateson has suggested that patterns of incest avoidance in humans and nonhuman animals can perhaps best be explained and understood if we postulate them to be the results of several biologically based capacities rather than of a single evolutionary capacity for incest avoidance, as sociobiologists like Wilson and their philosophical allies like Michael Ruse might have us believe. On the basis of both his own studies with quails and those of other researchers on the mating systems of other animals, Bateson contends that there is good evidence to believe that nature has acted to produce animals that have a tendency to choose mates that are similar but not too similar to themselves. The development of this capacity might be accounted for by the complex interaction of imprinting and habituation mechanisms. This mate-selection mechanism, he suggests, could be coupled in humans with at least two other capacities. The first is a tendency to conform to the behavior of the group with which one lives and to repress the behaviors of nonconformers. The second is a linguistic capacity. The mate-selection and conformism mechanisms could be linked, even before the ontogenetic or phylogenetic advent of linguistic capacity, to form an incest inhibition. With the development of language in the species and in the individual, this inhibition could be reinforced linguistically. Even this model is very simple, as the work of Kitcher (1990), who develops it further, demonstrates. But what it shows is that an attempt to establish the existence and nature of evolutionarily based moral capacities will more than likely require of us developmental analyses. Our moral sense will probably turn out to be a complex capacity that will need to be analyzed into its basic components. We will then need to determine the function of these components and figure out how they develop in an individual and evolve in a population.

Granting, then, that our moral sense is probably a system of individually complex, evolutionarily based moral capacities, let's focus on the evolutionary character of these complex capacities. Using our notion of natural selection as a selection theory, we know that we can distinguish abstractly an evolutionary capacity from a nonevolutionary one on the basis of the source of the variation for which there is selection. So we can say that both evolutionarily and nonevolutionarily based capacities may share these aspects: (1) they are variations; (2) they lead to differential survival and reproduction; and (3) critical environmental factors select for these variations. But evolutionarily and nonevolutionarily based capacities differ because the former are heritable and the latter are not. What this implies, of course, is that the former are genetically based and the latter are not. However, putting it this way is clearly inadequate, because we know that nonevolutionary capacities must themselves be based in some way on evolutionary capacities.

What is the difference between these two ways of being based on evolutionary capacities? What makes something a genetically based variation and something else a variation that, although ultimately based on evolutionary capacities, is itself nongenetically based? We can creep up on a distinction, though, given current limited understandings, it may still be difficult to glimpse it, even when we get up as close as we can. There are several ways of describing the basing relation that seem to be inadequate. First, we can say that genes *influence* phenotypic characters. But it is also true that genes influence nonevolutionary capacities as well. Our perceptual powers and cognitive powers are to some extent genetically based. For example, sensitivity levels of perceptual powers and processing speeds for problem solving are probably genetically influenced. If so, these genetically based effects can be expected to influence the responses, for instance, of a person who is determining what to do to help a bystander in distress. But we would not want to say that variations in response to bystander distress are genetically based. On the other end of the spectrum, we might say that when we have a genuine evolutionary capacity, then a gene or, more than likely, a number of genes alone *determine* the phenotypic variation. But that is clearly too strong since, despite what we may have often heard, *innateness*, to use an old-fashioned term, is not opposed to environmental influences like learning. Even a modification of this last version, namely, that for an evolutionary capacity genes are *necessary but not sufficient*, does not get us very far. It faces the same sort of problem as our first formula about influence.

We are tempted at this point to resort to a favorite current slogan: "It's all a matter of the interaction of genes and environment!" Behavioral geneticists have statistical resources for sorting out the respective contributions of each to the variation in the phenotype. Although I will not go into details here, there are good reasons for contending that this approach is limited (Feldman and Lewontin 1975; Lewontin 1974; Bateson 1987). Two limitations can be noted without getting into statistical details. Sometimes environmental and genetic contributions cannot be separated out since their respective contributions are not independent. To illustrate this point, Bateson shows how an adopted child's IQ can be the result not only of the separate contributions of biological parents and adoptive parents, but also of the difference between the child's genetically based IQ (as measured, for instance, by the performance of a biological parent) and those of her adoptive parents. The latter effect is one that is brought about by the interaction of environmental factors – the adoptive parents' intellectual influences – and genetic factors, the child's IQ. But, second, notice that a statistical analysis of the relative contributions of environment and genes, even if successful, does not enable us to sort out the operations of the actual causal mechanisms. We are not able to determine what, in fact, are the evolutionary capacities and distinguish them from the nonevolutionary capacities. At this point, we may again be tempted to follow the path of least resistance and resort to our developmental slogan. However, Bateson points out that our sloganeering, although understandable, is premature. He advises that we need to take the long, hard path of learning from actual developmental analyses. We have seen the sort of thing that he has suggested in regard to incest avoidance. But, as you might suspect, that is but the tip of the iceberg.

77

Developmental biologists are far from approximating, even with respect to morphological and physiological characteristics, Wilson's idealistic genetic basis hypothesis in which the pathways from genes to behavior are traced. Nevertheless, let's look at this suggestion more closely to garner some more general lessons about the traditional dichotomies between genes and environment and between innateness and learning. By doing this, we can come to an approximate account of an evolutionary capacity that we can use to explore our questions about evolutionarily based moral capacities.

Let us idealize the life cycle (Lewontin 1974; Sober 1984a). Assume that life is divided into two phases, developmental and adult learning.[1] The developmental phase begins with the zygote and ends with adult phenotypic characters, in our case, with some or all of the structural features of moral agency. The structural features are constituted by the four-level system of agency: a base level, including, if there be such, evolutionarily based moral capacities, along with the behavioral, reflective, and self-referential levels. As we shall see in more detail, we are thinking of three of these levels as constituted by various beliefs and desires. The adult learning phase begins with the adult features and builds from there. In explaining an action, using the indirect approach, we move from an explanation in terms of capacities that have been acquired during the adult learning phase, capacities that we have called nonevolutionary, to those possessed at the beginning of that phase, what we have called evolutionary. The former have been acquired during the adult learning phase in a variety of ways that we shall discuss in more detail later but are worth noting now. These include habituation and sensitization, respondent and operant conditioning, affective learning, and various sorts of representational learning, for example, perceptual, learning through modeling, symbolic learning, both verbal and nonverbal, and self-referential learning. Our general assumption is that the more cognitively or representationally complex the learning is, the more likely it is that the capacity constituted by it belongs to the adult learning phase.

Now let us consider the developmental learning phase in more detail, since it is there that we will find, if we are to find them, our elusive evolutionary capacities. We can begin with another slogan that proves to be inadequate: "Think of genes as coding for a phenotypic characteristic." Since that slogan leads us to think that there is some sort of one-to-one correspondence between a gene and the phenotypic character for which it is responsible, that slogan is inadequate. One way to see that such a one-to-one correspondence does not occur is simply to consider the amount of DNA in an organism and the size and complexity of the nervous system (Changeux 1985; Bateson 1987). For instance, it turns out, not unexpectedly, that mice have much more DNA than do bacteria; they also have a much larger and more complex nervous system than the corresponding "nervous system" of a bacterium. Yet, even though the size and complexity of the nervous systems of mouse and human differ substantially, they have about the same amount of DNA. This tells us, among other things, that the emergence of behavioral capacities in mammals was not achieved merely by the accumulation of genes or genetic materials. The development of behavioral capacities is dependent on the ways that genes and their products interact, the timing of their interactions, and the environments in which they interact. Thus, even though we assume that if selection is acting on something, so that there is some genetic product that regularly and reliably appears by the end of the developmental phase in our idealized life cycle, the process of its developmental emergence is not a simple genetic effect.

Consider an analogy used by Bateson (1986) to make a somewhat different point: a recipe for a cake. The words of the recipe stand for various ingredients and various processes that all go into making a cake. To start with, there's sugar and flour and baking powder and butter. But we do not expect each crumb of the finished product to correspond to an original constitutive component. We can make that assumption perhaps with the crushed walnuts and the cherry on top, but that's about it. Similarly, geneticists and developmental biologists tell us that the initial ingredients and the final product look very different from each other. Our final products are evolutionary capacities or the normal results of the developmental phase of the idealized life cycle. Our ingredients are the genes – more precisely, the alleles that make up each gene in a sexually reproducing organism and the environments in which they are situated. It is reasonable to assume that, for the most part, multiple alleles from different genes are involved in the construction of a single phenotypic characteristic. Thus the evolutionary capacity, that is, the effect, is a *polygenetic* one. In addition, we can assume that each allele makes contributions to the constitution of many different evolutionary capacities. We have what biologists call *pleiotropy*. As we have already mentioned, genes interact. That's called *epistasis*. And since we are dealing with sexually reproducing organisms, in reproduction certain genes are normally linked together and others are not, appropriately called *linkage*. But at the same time, in sexual reproduction we can expect a *recombination* of genes. So the zygote that starts the process will itself differ in each generation from its parents and usually from its siblings.

However, besides the complexity introduced by multiple genetic effects, there is also the complexity brought about by environmental effects. A portion of these effects of the environment in the developmental process involves learning. What counts as learning in the developmental process? This is not an easy question to answer. One common way to answer it is to consider learning as originating in the environment. But what constitutes the environment? Often we think of the environment as consisting of those entities and processes external to the developing organism. That implies that the maternal environment of the developing zygote must count as an environment. There is also some reason to move the environment within the organism since the action of the environment associated with learning is often contrasted with the actions of the genes. Let's do that and focus on the development of the brain. We can distinguish both the direct and indirect effects of genes in the construction of individual neurons and large assemblages of neurons, the effects of the so-called genetic envelope, from other environmental influences (Changeux 1983). The genetic envelope controls such processes as the division, migration, and differentiation of nerve cells, as well as the development of the so-called growth cone that guides the developing nerve cells to their proper locations. The genetic envelope also brings about the aggregation of appropriate cell types by enabling cells to "recognize" their "cell-mates." In addition, it fosters multiple connections, as well as determining the processes by which the neurotransmitters are assembled at the synapses. Finally, the genetic envelope is responsible for the spontaneous activity of the developing organism.

Given this account of direct and indirect effects of genes in the developing brain, where does learning conceived of as "environmental input" occur? Changeux distinguishes between causal processes and effects due to the genetic envelope and those that bring about the pattern of neural connections that are the final results of normal development. He contends that the stabilization of connections that result in the normal functioning of the brain is brought about by a *selection* process. This process involves a feedback mechanism from the postsynaptic cell. In normal development there is a profusion of presynaptic cells, all potentially able to connect with the postsynaptic cell. The feedback mechanism prunes away these excess connections, which are in a labile state. Changeux suggests that the mechanism is a selection mechanism whose operation is analogous to that of the physical environment in natural selection (Changeux 1983; Edelman 1987, 1992). Whether this is the case or not is not important for us. What is significant is that the sources of the activation of the feedback mechanism instrumental in the stabilization of the synaptic connections are initially the infant's spontaneous movement and the influences due to the mother. It is arguable whether the selective stabilization process should be called learning as opposed to, say, the construction of the neural substrate for learning. Fortunately, we can set that issue aside since it is part of a larger issue concerning the relationships between neurobiological and cognitive accounts of our mental capacities, including our moral capacities, that we discuss in Chapter 7. Here the point is that genes and environment interact very early in developmental processes that are important in the formation of capacities that are clearly relevant to and, perhaps, constitutive of evolutionarily based moral capacities.

Moving from the input of "environmental feedback" within the developing organism and from the maternal environmental influences to the external environment, we find a plethora of evidence on both the nonhuman animal level and the human level for the role of learning in normal development (Changeux 1983; Bloom and Lazerson 1988). For instance, studies have shown that sensory input is essential for proper development in a mouse's brain. Unless it gets feedback from whiskers on its snout, important neuronal groupings called *barrels* will not develop properly. Deprivation of visual input to rats, kittens, and monkeys also brings about corresponding abnormalities in their respective neuronal systems. Moreover, enriched laboratory environments – ones with large cages, many compartments, and playthings – have proved to be far superior to barren environments for the neuronal development of rats. Those in the former environment had a thicker cerebral cortex, more dendritic spines on cortical neurons, more dendritic branching, and more sites for the secretion of neurotransmitters. Whether this sort of environmental input with its resultant neurobiological effects counts as learning itself or as a precondition for learning, it seems clear that environmental inputs are a necessary part of normal development. They are part of what we have called the *developmental phase* of the ideal life cycle. We should then expect environmental inputs and/or learning to play a role in the ontogeny of evolutionarily based moral capacities. This conclusion fits in with our earlier considerations concerning the norm of reaction.

When we discussed the concept of norm of reaction in Chapter 3, we saw that we can expect phenotypic variation in a given genotype because of environmental variation. When we are considering mammals and humans, the environmental variation includes variations in the learning environment. Obvious candidates are parents, care givers, and siblings. So the normal and reliable result that we have postulated as the end product of the developmental phase will, when looked at from a population perspective, appear as a statistical distribution of some sort – perhaps, though not necessarily, a normal distribution. Bateson notes that if individual variations were merely the results of multiply influenced developmental processes, then we would expect a continuous distribution of a phenotypic character. But the distribution is not always continuous. Sometimes it is modal, that is, there are several segregated peaks with intervening valleys in the curve representing the measure of the phenotypic character. One way to interpret this discontinuous distribution is to postulate that it is due to "evolutionarily stable mixes of genetically distinct individuals" (Bateson 1987, p. 9). In other words, genetically different individuals have different behavioral and psychological capacities that are adaptive. Interpreted in game theoretical terms, this means that genetically different individuals have genetically based distinct behavioral strategies that are stable. An alternative hypothesis, preferred by Bateson, is that each individual possesses different genetically based strategies that it can employ, given different environmental conditions during the juvenile developmental phase. If Bateson is right, then evolutionary capacities will vary because of varying learning environments in the developmental phase. In addition, we can expect that these evolutionary capacities will themselves be complex, both in terms of being constituted by other evolutionary capacities, as in the case of our incest avoidance example, and in terms of being contextual; that is, their activation will depend on specific circumstantial and stimulus conditions, as well as on intrinsic conditions such as sensitivity or threshold levels.

Dawkins (1976) has argued that since selection operates on the phenotype, that is, on the organism after it has achieved its adult stage, one need not be concerned with all these developmental complexities. Bateson, however, points out that even though there is selection for the phenotype, it will not do to neglect development. For if we do not pay attention to the developmental phase, we are unable to determine the conditions that are responsible for helpful or harmful development. But when we do pay attention to the conditions of development, we realize that developmental environments differ. Moreover, it is naive to think that there is complete buffering against these environmental influences. The postulation of such buffering is what has led to the notion that evolutionary capacities are unlearned or inborn. Wilson attempted to remedy this view of his position in his work with Lumsden (1981) by characterizing the development of evolutionary capacities in terms of epigenetic rules. In their view, learning is involved in development, but it is constrained by rules of development. These rules are fixed and not learned. Bateson, however, believes that the situation of development is even more complicated. The learning rules, he contends, are themselves environmentally conditioned. He cites several striking examples (Bateson 1987). Many reptiles can develop either as males or as females, depending on the environmental conditions of their development. For example, if the eggs are below 30 degrees Centigrade at a certain critical moment in development, young turtles become females; if they are above this temperature, they become males. Male Gelada baboons can become either harem holders or small, female-like followers of the harem holders that achieve their copulations secretively. The former strategy is more adaptive if there are few harem holders, the latter if there are many. Thus, from a developmental point of view, it is important for the young male to make his "decision" at the latest possible point and with the best environmental information.

What then can be said about an evolutionary capacity and how it differs from a nonevolutionary capacity? In terms of our idealized life cycle, we postulate that evolutionary capacities are the end products of the developmental phase of the life cycle. These products are the result of multiple genetic influences, one allele affecting many evolutionary capacities and many alleles affecting one evolutionary capacity. Moreover, genes exert their influences by affecting the actions of other genes. In addition, both evolutionary and nonevolutionary capacities are environmentally influenced and, in the cases we are particularly interested in, those of humans, environmental influence sometimes takes the form of learning. Development does not exclude learning. Nor is the learning and its results fixed in any straightforward manner. The products of the developmental phase are what natural selection selects for and against. These are the complex results of genes, environments, and their multiple interactions. Subsequent variations are subject to cultural selection whose "ends" sometimes coincide or overlap with those of natural selection and at other times are neutral to those of natural selection or even in opposition to them.[2] We anticipate that evolutionarily based moral capacities as evolutionary capacities will more or less fit this abstract and sketchy pattern. But, keeping in mind one of Bateson's major points, we should expect to find out much of evolutionarily based moral capacities and the moral sense by looking at developmental details. It is to those details that we now turn.

4.2 MORAL DEVELOPMENT AND THE MORAL SENSE

During a good portion of the twentieth century, the study of moral development in children has been dominated by three very different traditions in psychology: Freudian, behaviorist, and cognitive developmental. I shall have more to say about these traditions in Chapters 5 and 6, but I need to say a few things at this point to set the stage for our discussion of Martin Hoffman's views on moral development and empathy. In their own way, both Freudians and behaviorists emphasized the noncognitive and nonconscious bases of our moral agency. In different ways, both traditions viewed morality as the imposition of an external agent. Jean Piaget and Lawrence Kohlberg, the two giants of the cognitive developmental tradition, and their followers emphasized the rational and conscious character of moral agency and argued that moral agency is the natural result of human development under the influence of proper environmental stimulation. The theories and approaches of all these traditions have undergone substantial change, but it is fair to say that the cognitive developmental views had become prominent by the 1960s. It is easy to see that the cognitive developmental view has the advantage of corresponding more closely to our reflective, commonsense views about the nature of our moral agency. In addition, it seemed to have some substantial empirical support. Because in their view moral agency is essentially a cognitive, conscious, and rational activity, cognitive developmentalists focused their studies on children and adolescents rather than on infants and toddlers. They did not expect that morality or moral agency would be found in infants and very young children, those younger than five or six years of age. A number of factors, however, have led developmental psychologists to the study of moral development in infants and very young children. The results have been truly surprising and, for those interested in the issue of moral sense and our possession of evolutionarily based moral capacities, very exciting.

I consider the work of one prominent developmental psychologist, Martin Hoffman, who has worked out a theory of moral development that focuses on the central role of empathy in moral agency. His emphasis on empathy in moral development is especially interesting for several reasons. First, there is a long philosophical tradition that has looked to the emotions and our cognitive/affective capacities as the sources of our moral agency (Bradie 1994). David Hume (1946, 1957) comes to mind immediately in this regard. Second, Charles Darwin (1871) gave pride of place to empathy in his own account of the development of what he called the *moral sense.* Third, there seems to be growing evidence of an evolutionary basis for basic emotions such as happiness, sadness, disgust, anger, surprise, fear, and interest (Campos et al. 1983). Fourth, empathy seems to be an important source of altruistic behavior.

Although I focus on empathy as an evolutionarily based moral capacity, we should keep in mind that the moral sense is more than likely a complex system of a number of evolutionarily based moral capacities, and that the base-level system of moral agency is composed not only of evolutionarily based moral capacities but also of learned, nonevolutionarily based moral capacities. Evolutionarily based moral capacities may be relatively general, for instance, those responsible for feelings of guilt or anger, or they may be more specific, for instance, the incest avoidance mechanism postulated by Bateson discussed earlier. Indeed, there is recent interesting, scientifically informed theorizing concerning the phylogenetic origin of rather specific capacities that might either be evolutionarily based moral capacities or morally relevant evolutionary capacities, for instance, capacities and propensities concerning mating, parental care, and sharing (Barkow et al. 1992). I focus, then, on empathy, specifically empathic distress, as a model system for exhibiting the ontogeny of an evolutionarily based moral capacity. Finally, I briefly examine some of the evidence for the evolutionary origin of empathic distress.

There are theoretical differences about how to define empathy (Barnett 1987). One important one centers on the cognitive/affective mix in the empathic response. Some theorists understand empathy to be primarily a cognitive ability to recognize and understand the thoughts, perspective, and feelings of another individual. Others conceive of empathy as primarily an affective ability. This theoretical issue not only raises an important problem about the nature of empathy but also raises a more fundamental question about the distinction between cognition and affect. Without attempting to answer that question at this point, it seems reasonable to say that both cognitive and affective processes are representational, but that such cognitive processes as reasoning and deciding have more representational power than do such affective processes as emotions. If that's so, then the understanding of empathy as primarily an affective process is better suited for discussions of development. Following a number of developmental psychologists, including Hoffman, we can take empathy to be an affective response more appropriate to someone else's situation than to one's own. Empathic distress is a response that is more appropriate to the situation of the person in distress than to the empathizer who is not in the situation. Empathy needs to be distinguished from other states with which it might be confused. Empathy is a state of the empathizer. It is not the same thing as merely having the same or similar state as another. Nor is it merely having the same or similar state as another because that person has that state. It is also having an *emotional* state that is the same or similar to that of the person with whom one is empathizing *because* that person has that state. I take an emotional state to be a complex one with the following characteristics: (1) representation of something as desirable or undesirable; (2) feeling, interpreted as hedonic tone; (3) bodily sensations; (4) involuntary bodily responses and overt expressions; (5) tendencies to act; and, sometimes, (6) an upset or disturbed condition of body or mind (Alston 1967; Campos et al. 1983). I omit the complication that a person in distress may not always have the appropriate psychological state for a distressful situation. She may, for example, have been rendered unconscious by a serious accident. Moreover, mere cognition of another's state as desirable or undesirable is not sufficient for empathic distress. As a result, what occurs in role taking or perspective taking may not be empathy since these need not be emotional states. I consider having the same (or similar) emotional state as another because of the other's state to be the key element in empathy. Sympathy adds to that the condition of concern for the other. Empathy and sympathy, as we shall see, often result in altruistic tendencies and actions. By *altruistic* I mean what the psychologists often intend by the term *prosocial*. Prosocial intentions and actions are those that have the benefit of another as their object. This leaves open the degree of self-regard in the actions and intentions.

4.3 MARTIN HOFFMAN'S ACCOUNT OF MORAL DEVELOPMENT

Using the account of empathic distress that we have just considered, Martin Hoffman (1984a,b, 1988) argues that there are a series of developmental levels of empathy. He does not intend the notion of levels in the strong sense used by Kohlberg or Piaget; rather, he means that a pattern in the development of empathy can be discerned and that these levels involve greater cognitive, affective, and behavioral complexity.[3] Hoffman distinguishes four levels of empathic development: (1) global, (2) psychologically undifferentiated, (3) immediate, and (4) extended.[4]

Global empathy occurs in children less than one year of age. One of its precursors, in Hoffman's view, is the phenomenon of neonate responsive crying. This is the highly confirmed phenomenon that newborn infants cry in response to the crying of other infants. Although there is some question in the literature about whether neonate responsive crying ought to be classified as a precursor of empathy, it does seem to be a distinctive response, not a startle reaction or a reaction to aversive noises generally. Global empathy, the first level of empathic development, is a response to another's pain as if the person responding is herself in pain. Hoffman postulates that at this stage the infant has not distinguished herself from other selves. There is no object permanence of other persons. So the painful state of the other is also the painful state of the infant. We might also postulate that there are behavioral signs of recognition of a departure from a standard, but it is not at all clear what kind of cognitive state corresponds to such behavioral signs. There is no evidence of any assessment of the source of the distress in the other. And, of course, there is no helping response.

Undifferentiated empathy, the second level of empathic development, occurs in one- and two-year-old infants. They have achieved the notion of person permanence but have not yet achieved the understanding that the inner states (the thoughts and feelings) of the other person are independent of their own. So they respond to the other's distress by doing for the other what would be helpful to themselves. For instance, a child may bring her own mother to help another child in distress even though the latter's mother is present. Thus there is evidence of a cognitive grasp of the other as distinct from the self and a recognition that the other has psychological states, although not that they can be different from one's own. The child also seems to grasp in some manner that there is a harm, a departure from the standard or what ought to be, but it is not clear what this understanding amounts to. Nor does the child seem to understand at this point in her development what the source of the harm is. There is some evidence of a tendency to help, although the helping may not be appropriate. By the time a child is about three years old, she recognizes the distinction in psychological states and begins to offer appropriate help. This is the level that I have called immediate empathy.

In later years, empathy is extended in the sense that the person is recognized to have experiences beyond the immediate situation and that these too become the bases for an empathic response. Indeed, the person's general condition can invoke empathy. And empathy can extend beyond an individual to an entire group. Thus the ontogeny of empathy seems to reveal patterns of generalization with respect to the conditions of an individual person and from individuals to groups of individuals.

It is important at this point to mention some emotional reactions that are closely connected with empathy and sympathy. Hoffman notes that empathy or sympathy can turn to personal distress in some situations. In those circumstances the focus of the empathizer's concern shifts to herself, and the empathizer escapes the stressful situation in order to achieve some relief or takes action on behalf of the other in order to give herself some relief. If the source of the other's distress is the empathizer, empathy and sympathy can be accompanied by guilt, the assessment by the empathizer that she herself is the source of the person's distress. On the other hand, if someone else is judged to be the source of the other's distress, anger toward that person can accompany empathy and sympathy. In both of these latter situations there seem to be implicit assessments that the person in distress is a victim, that is, that she is not responsible for her own distress.

4.4 AN ASSESSMENT OF HOFFMAN'S HYPOTHESIS ON THE DEVELOPMENT OF EMPATHY

The empirical literature provides support for Hoffman's hypothesis on the development of empathy. First, let's consider what can be taken to be rudimentary forms of empathy or its precursors. Three phenomena are of particular interest in this regard: (1) reactive crying in neonates; (2) affective synchrony in mother (or care giver)–infant play, which begins when the infant is about 2–3 months old; and (3) social referencing, which begins when the infant is about 10–12 months old. A brief examination of each of these will help us understand the development of the infant's abilities, a development that culminates in genuine empathetic responding by the time she is 30–36 months of age.

Newborn infants seem to cry in reaction to the crying of other infants, especially the crying of infants of approximately the same age (Simner 1971; Sagi and Hoffman 1976; Martin and Clark 1982; all cited in Thompson 1987). Simner found that neonatal crying produced reactive crying in other newborns significantly more than did background noise, the cry of a 5-month-old, or a synthetically produced cry. Hoffman and Sagi explain this phenomenon as a rudimentary empathic distress reaction. Hoffman also offered an alternative conditioning explanation: The reactive crying is conditioned by the previous occurrence of the other infant's crying before a nonreactive crying episode of the infant in question (Hoffman 1982). Thus the other infant's crying became a conditioned stimulus for the first infant's crying. Piaget, on the other hand, thought that reactive crying could be understood as the assimilation and reproduction of a piece of behavior (Piaget 1951, cited in Thompson 1987). This last interpretation presupposes a similarity between the originative and reactive crying that appears not to have been found (Martin and Clark 1982, cited in Thompson 1987). Ross Thompson understands reactive crying to be a kind of emotional contagion evoked in part by the general similarity of the crying to the reactive neonate's own signal of distress. Since neonate reactive crying is not likely to be a response to the distressful situation of the other, Thompson does not believe that neonate reactive crying is a form of empathic response. He also points out that a similar form of emotional contagion seems also to be present in somewhat older infants (Thompson 1987). By the term *contagion* he seems to have in mind a widespread, noncognitively based imitative behavior.

By two or three months of age, a second precursor of empathy begins to appear. Care giver and child begin to engage in an exchange that displays an affective synchrony. Developmental psychologists refer to the care giver as "mother" since the mother has traditionally been the primary care giver and since affective synchrony has been studied with mothers and their babies. Each begins to respond to the expression of the other, often in a playful exchange in which the smile or excited expression of the one is greeted with a similar or even heightened expression by the other. The ability of the infant to engage in this sort of exchange depends in part on her ability to scan the mother's face visually in an organized fashion. Beginning at about two months, the infant starts to discriminate among facial expressions of happiness, anger, surprise, and other emotions. She seems to have achieved this ability by about five months of age. At this point it is not necessary to postulate that the infant attributes any emotional meaning to the different facial expressions. That is to say, the infant does not yet possess the ability to discern that the other has psychological states, in this case emotional states. However, that ability soon flowers.

By the time the infant is 10–12 months old, she understands that emotional expressions have emotional meaning, that they reveal the psychological state of the care giver. The evidence for this is in the striking phenomenon known as *social referencing*. Infants at this age actively try to obtain emotional cues from others, especially their mothers, in order to assist them in their own evaluation of uncertain situations. One important source of support for social referencing is the visual cliff studies. In a visual cliff study, the infant is placed on a setup that makes it appear that she is near a sheer dropoff. The mother is at the other end of the apparatus. The child looks to her mother and searches for the emotional cues that will enable her to determine whether it is safe to proceed. Whether the infant will go to her mother depends on the emotional cues the mother provides. If the mother displays a positive emotional state, the infant moves across the cliff to her. If her expression is negative or neutral, the infant does not. The claim that by about the end of his first year an infant is able to grasp that another has psychological states is also supported by studies of early referential communication and linguistic and protolinguistic development. Trevarthen (1982, cited in Thompson 1987) has argued that infants as young as 12 months of age attempt to achieve a commonality of subjective states with another. They do this when (1) they try to direct another's attention to something by, for instance, pointing at an object, looking at it, or vocalizing; (2) they attempt to acquire assistance by, for example, tugging at their mother's hand; (3) they attempt to obey simple requests; and (4) they try to return affectionate gestures. By these actions the infant seems to be exhibiting that she understands that the subjective states of an individual differ and have the potential to be shared. Trevarthen claims that this understanding that a joint subjective state can be shared provokes an interest in achieving this shared or mutual subjectivity. That in turn, he believes, provides the basis for the development of a variety of communicative and cooperative behaviors in subsequent months. Shortly after the appearance of the ability to discern that another has psychological states, in particular, to discern that another's expression can reveal the emotional state of that person, infants begin to display genuine empathic responses including the tendency to help. By 30–36 months of age this ability seems to be firmly established.

Alternative interpretations of infants' reactive distress and attentive concern have been suggested. Some researchers have suggested that the reactive distress of the infant may be due to a startle response or to self-concern. Others have claimed that the attentive concern of the infant can be explained as an orienting response or as efforts to interpret and understand another's emotional behavior rather than as empathy. Moreover, some infants ignore the distress of others, and there is a good deal of individual variation. Thus there is some reason to question the claims about the presence of empathy in infants. Thompson, nevertheless, supports the claim that infants are capable of an empathic response because of both the connection between their emotional response and prosocial initiatives and the connection between an infant's punishment of herself and another's distress, whether or not the infant is the source of the other's distress (Thompson 1987). This tendency is increased by the mother's showing the infant that she has caused distress and that she ought to do something to repair the damage. If the infant's response were primarily an orienting, startle, or initial avoidance response, it would not be accompanied, as it often is, with attempts to help the person in distress. Thus, because of its significance for Hoffman's theory and for the account of empathy that we have been discussing, we should look in more detail at the evidence of the connection between empathic emotional expressions and helping responses.

Marian Radke-Yarrow, Carolyn Zahn-Waxler, and Michael Chapman (1983) present a comprehensive discussion of the development of empathy and prosocial behavior in children, including infants and toddlers. Their studies and those of others indicate that before three years of age, infants display prosocial behaviors. In one study, researchers have found that infants 15 to 18 months of age share, cooperate, and give care in situations that do not involve stimuli indicating acute stress (Rheingold, Hay, and West 1976; Hay 1979; Rheingold 1979; all cited in Radke-Yarrow et al. 1983). Studies of 12- to 24-month-old infants provide evidence of increased cooperative behaviors (Eckerman, Whatley, and Kutz, 1975; Ross and Goldman 1977; both cited in Radke-Yarrow et al. 1983). Rheingold also contends that many behaviors of young children that had been interpreted as manifestations of dependency, approval seeking, and attention seeking are instead indications of the children's ability to give and to share with others (Rheingold et al. 1976; Hay, 1979; Rheingold 1979; all cited in Radke-Yarrow et al. 1983). She has in mind such behaviors as showing and giving objects to other people, behaviors that infants display beginning at around their first birthday. Zahn-Waxler and Radke-Yarrow (1982) have studied extensively the helping response of infants in situations of acute stress. They examined groups of infants aged 10, 15, and 20 months in naturalistic settings for a period of 9 months. Mothers were trained to give detailed observational reports of naturally occurring distress incidents and their infants' reactions. In addition, both investigators and mothers simulated mild emotional distress. Their findings confirm Hoffman's theory of empathic development. At 10–12 months of age in one-half of the incidents of distress, infants respond to a situation of distress with a frown, a sad face, crying, or visual checking with their care givers. At 16 months of age, the infants' general agitation is less pronounced, and their concerned attending becomes more prominent. In addition, positive attempts to help the person in distress begin to appear. For instance, the infants begin to pat or touch the person in distress. By 18–24 months of age, these positive attempts to help increase and become more differentiated. Infants at 24 months of age bring objects to the person in distress, make suggestions about what to do, express sympathy verbally, bring someone to help, protect the victim, and attempt to make the person feel differently. Moreover, they try something else if their initial attempts do not work. This seems to indicate that the infants can keep the distress of another person in mind and view it as a problem to be solved. It must be pointed out that infants are not always responsive to the distress of another, and sometimes they try to escape or avoid the distressing stimuli or even attack it. But these responses seem to be much less frequent than the positive, helping reactions. Dunn and Kendrick report similar findings on the early appearance of comforting responses (Dunn and Kendrick 1979, cited in Radke-Yarrow et al. 1983). And they report somewhat similar findings on empathic responses in an extensive longitudinal study of early sibling relationships. However, Mahler, Pine, and Bergman report non-uniform responses (Mahler, Pine, and Bergman 1975, cited in Radke-Yarrow et al. 1983). Radke-Yarrow and Zahn-Waxler (1982) conclude that before infants are three years old, they are capable of and exhibit empathic prosocial behaviors.

Studies and observations in naturalistic settings of two- to six-year-olds indicate that very young children are not only capable of egocentric, selfish, and aggressive behavior but also have the capacity to form affectionate relationships and display prosocial behaviors in a variety of settings and with a variety of persons. Barnett (1987) points to a number of factors that have been shown to have a positive effect on the development of empathy in children. These are secure and early attachment, parental affection, availability of empathic models, use of parenting techniques that show the child exactly what the effects of her or his actions are, encouraging the perception of similarity to others, discouraging excessive interpersonal competition, and encouraging the development of a positive self-concept. According to Barnett, even though current research indicates that the connection between any one of these factors and empathy, including prosocial responses, is weak, it nevertheless seems to be the case that empathy and related responses are best brought about by social environments in which (1) the child's own emotional needs are satisfied and excessive self-concern is discouraged, resulting in a situation in which the needs and emotions of others become more salient for the child; (2) the child is encouraged to experience, identify, and express a wide range of emotions; and (3) the child is provided with many opportunities to experience and interact with others who model empathy and prosocial responses.

Studies of 6- to 16-year-olds provide little data for developmental claims. Claims are often made that children's prosocial behavior increases with age. However, little attention is paid to the types of prosocial behavior displayed, and development seems to be measured merely in terms of increase in frequency. But Radke-Yarrow, Zahn-Waxler, and Chapman (1983), in their survey of the literature, note that variations in the frequency of prosocial behavior depend on the type of prosocial behavior, the ages of the children involved, and the research methods employed. Thus, they conclude that the data do not allow any conclusions to be drawn about the development of prosocial behavior during this age period.

One further problem needs to be mentioned. It has often been noted that, in comparison to the connection found in adults, there seems to be little or no connection between empathic response and prosocial behavior in children. Since according to Hoffman's theory of empathy that connection is key in determining the presence of genuine empathy, we need to consider the implications of these findings. It is clear that apparently altruistic behavior can be motivated by nonsympathetic, pragmatic, or selfish reasons. In addition, it is clear that genuinely altruistic behavior can be motivated by something other than sympathy, for instance, moral values and principles. How do we explain the apparent differences in children and adults? Nancy Eisenberg and Paul Miller (1987) suggest several factors that may account for these differences that do not entail the rejection of a link between infants' and children's prosocial behavior and their empathic responses. Psychologists make a distinction between *state empathy* and *trait empathy*. The latter is an enduring, cross-situational disposition; the former is an occurrent state. It is likely that there is a stronger connection between state empathy and prosocial behavior than between trait empathy and prosocial behavior. Although Eisenberg and Miller do not indicate why this is so, one might speculate that the immediacy of the stimuli in cases of state empathy and the overall lack of evidence for traits both argue for the greater strength of the connection between state empathy and prosocial behavior.

Eisenberg and Miller (1987) point out that there is some reason to believe that in studying the relationships between empathy and prosocial behavior in children, researchers have been measuring the connection between trait empathy and prosocial behavior. However, in their studies of adults, they have been attending to the connections between both trait and state empathy and prosocial behavior. Thus, they may not have been measuring the same phenomenon in children and adults. Researchers have used the following measures of empathic response: (1) picture-story indices, (2) question-naires, (3) self-reports of reactions in experimental situations, (4) reports of others, (5) facial/gestural indices, (6) physiological indices, or (7) experimental inductions of empathy. A survey of the studies shows that in adults, empathy as measured by a variety of these indices is positively associated with prosocial behaviors. The results with children, however, have been mixed. When experimenters used indices that require the self-report of the children, such as picture-story and questionnaire indices, they did not find consistent relationships between empathy and prosocial behaviors. But in the case of indices that did not require the ability to report about one's emotional state, such as facial and gestural indicators, there was a somewhat more consistent relationship between empathy and prosocial behavior. There are reasons to question the validity of picture stories as measures of empathy in children, and both picture-story and questionnaire indices are problematic because of the limited ability of infants to report their emotional reactions. In addition, in the adult studies, empathy and prosocial behavior have been assessed in the same situation. The potential object of the empathy was also the potential object of the prosocial behavior. This has not been the case with children. Empathy has been measured in one situation, and then a link with prosocial behavior in another situation has been sought. Thus researchers have been attempting to find a link between trait empathy and prosocial behavior rather than between state empathy and prosocial behavior. When they try to connect state empathy with prosocial behavior, the results indicate a positive connection, as is the case with adults. Of course, the differences between children and adults may be real and due to developmental disparities. These differences could be due to the greater difficulty that children have in understanding situations and determining what to do, as well as the greater difficulty of being able to do anything. Older children may be more inhibited in helping because their helping skills are less competent. In addition, because children are less capable of assuming the role or perspective of another, they may feel more personal distress and thus attempt to escape or avoid more often than do

93

adults. Eisenberg and Miller call for more and better research to clarify some of these issues.

Thus Hoffman's descriptions of the developmental situation seem to be supported in the literature. But his account is certainly not established; it remains a working hypothesis about the ontogeny of empathy, sympathy, and prosocial/altruistic behavior. Given this account, what can we say about the claim that empathy is a manifestation of a moral sense?

4.5   EMPATHY AS A MANIFESTATION OF THE MORAL SENSE

In order to establish the claim that empathy is a manifestation of moral sense, we need to establish its relevance to moral agency. To do this, we need to show that it meets satisfactorily substantive and functional criteria for relevance. In addition, we need to show that empathy is an evolutionary capacity, a component of an evolutionary capacity, or a set of evolutionary capacities. With respect to the relevance of empathy to moral agency, I have already suggested that empathy can be considered one of the first-level cognitive, affective, and behavioral dispositions that constitutes the first level of a four-level system of moral agency. Action based on a first-level disposition, like empathy, need not be free but can be incorporated in a free action. If empathy is an evolutionary capacity, it derives from the evolutionary "ends" of natural selection. These ends, of course, need not be moral, but they are not necessarily amoral or immoral either. Indeed, in the case of empathy, the proximate ends seem to be altruistic action since, as we have seen, there is evidence that empathy is causally related to altruistic behavior. Consequently, the substantive criteria of moral agency are also met by empathy. So I conclude that there is enough evidence to support the claim that empathy is a moral disposition. But is it an evolutionary capacity?

To help answer that question, we can examine and comment on Hoffman's (1981) own argument for empathy as an evolutionary capacity. Hoffman finds several lines of converging evidence for empathy as an evolutionary capacity. First, he argues that although the necessary cross-cultural studies have not been made with respect to empathy, the fact that it is found at all ages is indicative of a genetic component. Humans of all ages display empathic responses physiologically, by facial expression and verbal report. The import of this evidence seems to be that if empathy were merely the result of learning and cultural transmission, we might not expect to find it at all ages, especially in the very young. Of course, young infants learn an awful lot, so they may learn empathy at an early age. But, as we have seen, learning is a part of normal ontogeny. Thus its presence does not mean that

empathy is not an evolutionary capacity. Second, Hoffman argues that empathy is often an involuntary response. As such, it is not a reflectively
cognitive response to situations of distress, for instance. That may be indicative of its primarily genetic roots. However, involuntary responses are
not all necessarily due to evolutionary capacities. Primarily learned capacities can be exercised in an involuntary fashion. Nevertheless, its non-
reflective character and its early appearance in infants who have not undergone a lot of socialization give some plausibility to the claim that empathy
may be an evolutionary capacity. Third, Hoffman argues that there is evidence that empathic responses are based in the limbic system that we share
with other animals. So we might take this as an indication that our empathic
responses are evolutionarily ancient and primarily evolutionary capacities.

Perhaps, Hoffman's strongest argument for empathy as an evolutionary
capacity is his fourth argument: that the modes of empathic arousal are
indicative of an evolutionary capacity. Hoffman has identified four modes
of empathetic arousal[5]: (1) mimicry, due to automatic imitation and afferent
feedback; (2) conditioning and direct association; (3) language-mediated
association; and (4) putting oneself in the place of another. These four
modes of arousal are all present in adult behavior, although the last two
develop later, with the acquisition of linguistic capacities and role-taking
and perspective-taking capacities. Two of these, those that are automatic
and involuntary, provide evidence, though not conclusively so, for empathy
as an evolutionary capacity. The mimicry hypothesis is an old one but has
been recently revived and supported with suggestive evidence (Bavelas et
al. 1987). The idea is that humans spontaneously imitate the postures and
facial expressions of others. This imitation creates inner cues that, through
afferent feedback, allow one to understand and experience another's affect.
The conditioning mode can be explained through classical conditioning.
Suppose that a child falls and cuts herself. She feels pain and begins to cry.
The falling and the blood from the cut could become associated with pain,
distress, and crying behavior. When the child then sees another child in a
situation where she might fall or has fallen or is cut and bleeding, these cues
may evoke vicarious distress in her. One way to elaborate on the conditioning mode of arousal, an elaboration that applies to both respondent and
operant conditioning, is that the responses to the primary stimuli are built in.
Hoffman believes that such empathic responses are probably species-wide,
since there is probably a commonality of excitatory situations in human
socialization processes, including distress-provoking situations, and since
humans, because they possess a similar nervous system, could be expected
to respond to similar situations in similar fashion. Hoffman offers one final
argument for empathy as an evolutionary capacity. He suggests that the

developmental pattern fits the pattern of an evolutionary capacity that is open to cognitive and behavioral development. In such a pattern, the motive for a behavior appears long before the cognitive and coping skills necessary for engaging in the appropriate behavior have developed. This is the case with empathy. Hoffman's arguments, then, provide some support for the claim that empathy, in particular empathic distress, is an evolutionary capacity. Is there any further support for this claim?

We have seen that nonevolutionary capacities can mimic evolutionary capacities. They too can vary in a population and give rise in given environments to differential survival and reproduction. The crucial issue is whether empathy has a genetic component. Of course, a definitively positive answer depends on actually tracing the dispositional complex from its genetic bases. Doing that would give us a positive outcome to what we have called Wilson's *genetic basis hypothesis*. Given the lack of the best possible proof for a genetic connection, proponents of a moral sense have to rely on other, much less definitive hypotheses. Chief among them are, as we have seen, (1) the argument from cross-cultural commonality and (2) the animal comparison hypothesis. Of course, it is possible that the common dispositions are the result of cultural commonalities, so the fact that there is cross-cultural generality with respect to, say, empathic behavior and altruistic behavior, if it is a fact, is also explainable in terms of learned/culturally transmitted, nonevolutionary capacities. Moreover, animals also learn, and there is evidence for cultural transmission in animals. But neither of these problems is insuperable.

Let's consider briefly some support for empathy as an evolutionary capacity arising from each of these hypotheses. Franz de Waal (1996) understands morality naturalistically within the larger context of trying to explain the social behavior of primates, viewing morality as an evolutionary means for attaining social integration. He uses a wealth of anecdotal, observational, and experimental findings and combines them with theoretical hypotheses about psychological enabling mechanisms, including empathy, to make a strong case for the evolutionary continuity of moral capacities in humans and their nearest primate relatives (the chimpanzees, bonobos, gorillas, and orangutans, as well as both Old World and New World monkeys).[6] Robert Plutchik (1987) has suggested that animals display many different sorts of empathic behaviors, for instance, grouping behaviors, like schooling in fish, flocking and mobbing in birds, and herding behavior in mammals. Further examples are imitation, play, and bonding behaviors. Plutchik seems to be pointing to the presence of emotional communication in the animal kingdom. Such communication can involve the evoking of a similar emotion in another animal but need not do so. And even if it does invoke a similar emotion, that emotion need not be more appropriate to that of the object of empathy than to that of the empathizer. Nevertheless, some displays may result in the latter sort of empathy, for instance, distress displays. Moreover, the broad sort of empathic behavior that Plutchik discusses shows an evolutionary continuity with the narrower sort we have been discussing. Thus, he provides some support for the animal comparison hypothesis as applied to the thesis that empathy has an evolutionary basis.

Equally important for the claim that empathy is an evolutionary capacity, or has an evolutionary capacity as a component, is the strong evidence for the cross-cultural universality of expression and recognition of happiness, sadness, disgust, and anger, as well as suggestive evidence for surprise, fear, and interest (Campos et al. 1983; Izard and Saxton 1988). If it is the case that emotional communication has an evolutionary basis, that fact provides some support for the claim that empathy is an evolutionary capacity or has an evolutionary capacity as a component. For it implies that the emotional communication involved in empathic distress has an evolutionary basis. What is left open is the nature of the recognition of the emotion. It does not seem to follow from the fact of emotional communication that the recognition must involve the recognizer's experiencing the same or somewhat similar *emotion* as that of the sender. Nevertheless, it seems more likely than not that the evolutionarily based recognition mechanism involves emotional factors rather than more complex cognitive processes. Finally, we can add the role that emotional expression plays in the motivation of behavior.

Thus, although the case has not been definitively made, I believe that enough evidence already exists to make empathy a plausible evolutionary capacity or to have a component that is an evolutionary capacity.

4.6 THE MORAL SENSE AND MORAL AGENCY

I have been exploring one causal and explanatory connection between our biological and moral natures, arguing for the claim that humans possess an evolutionarily based moral capacity. This capacity functions, along with other capacities, as motivational bases for the exercise of moral agency. I believe that a plausible case has been made that empathy, specifically the capacity for empathic distress, meets the requirements for such a moral disposition. Of course, it need not be and probably is not, our only moral disposition. It seems clear that our moral sense is probably a complex disposition or set of complex dispositions.

Although, plausibly, some of our base-level moral capacities are evolutionarily based, others are probably learned and nonevolutionary. To ex-

97

plore this possibility, and to deepen and enlarge our understanding of the
nature of our moral agency and how it is acquired and put to work, I shall
examine the proposals of B. F. Skinner about our moral agency. Skinner is,
of course, Mr. Behavioral Psychologist himself, and a leading developer
and proponent of radical behaviorism. In the next chapter, I consider and
evaluate Skinner's contention that the science of operant conditioning is in
fact a science of values to determine what contribution his views make to an
integrationist account of moral agency.

NOTES

1. This idealization may be wildly wrong with respect to its temporal divisions and
   its assumptions about what is selected for if those developmental psychologists
   who extend the developmental process through old age are correct and if those
   developmental psychologists and biologists who support the thesis that there is
   selection for developmental systems are proven correct (Griffiths and Gray
   1994).
2. This is a major oversimplification, since we would expect to find cultural influences on the phenotype during the developmental phase before the attainment of
   adulthood and selective forces, for example, for various reproductive strategies
   and strategies for parental care, acting in the adult phase.
3. For more on Kohlberg's views about stages of moral development, see Chapter 6.
4. I have used my own descriptive terms except in the case of (1).
5. Hoffman actually discusses five modes, since he includes the mechanism for
   neonate responsive crying, which he considers to be a precursor of empathic
   behavior.
6. He identifies five general prerequisites for morality that have been shaped by
   natural selection; these are capacities for (1) sympathy, (2) mutual aid, (3)
   discerning and being motivated by a sense of fairness, (4) developing social
   norms and enforcing them, and (5) conflict resolution.

III

The psychological bases of moral agency

5

Behaviorally based moral capacities

How do we learn to behave morally?

5.1 MORAL AGENCY AND THE BEHAVIORIST CONNECTION

B. F. Skinner (1971) tells us that the science of operant behavior is the science of values. Using this simple and elegant proposal, we may have a way to link sociobiologically based theories about values with a theory of learning to form an integrated biologically and psychologically based theory of moral agency. The basic reinforcers on which all operant learning builds can be considered from the sociobiological point of view to be evolutionarily based. On the foundation of these basic reinforcers, we can then learn new and complex behaviors that enable us to achieve our goals in complex and changing environments. Can the behaviorist connection provide the necessary supplement to a sociobiological account of moral agency, making it an adequate account of our moral agency? In this chapter, I lay out the major points of Skinner's behaviorist account of moral agency. My goal is to understand and assess Skinner's claim that the science of operant behavior is the science of values and to determine to what extent our moral agency can be accounted for in terms of evolutionarily and operantly based moral capacities. Before we examine the details of Skinner's account of moral agency, let's sketch out some of the distinctive features of both behaviorism and Skinner's own brand of behaviorism.

Behaviorism is the movement in psychology that sprang up in the early part of the twentieth century and flowered around midcentury. It claimed, paradoxically, that if psychology was to make genuine advances as a true science of the mind, it had to abandon the mind and look to behavior! Behaviorists were impressed by the lack of success of psychologists at the end of the nineteenth century and the beginning of the twentieth century who attempted to study mental phenomena by means of careful introspection. In place of introspection, behaviorists thought that psychologists

101

needed to adopt the empirical methods of the natural sciences, especially those of physics. They were also convinced that Cartesian dualism, the view that humans are composed of two radically different sorts of entities, an immaterial mind and a material body, was untenable. Thus, they rejected explanations that invoked the mind as an immaterial factor that caused the person to do what she did. They considered these sorts of explanations to be empty and completely unscientific.

Skinner's own position is often described as modified radical behaviorism. Radical behaviorists advocate the avoidance of mental entities because they don't think there are any. Radical behaviorism is frequently contrasted with methodological behaviorism. The methodological behaviorist believes that, whether there are mental entities or not, proper scientific method in psychology demands the avoidance of any unobservable explanatory factors like mental entities. One way that this can be done is to employ operational definitions of mental factors. An operational definition is one in which a term that refers to some nonpublicly observable event or factor is defined in terms of publicly observable factors and events. Skinner agreed with the methodological behaviorists' admonition to avoid explanatory mental entities and held a modified radical behaviorist position. Unlike the pure radical behaviorist, he was willing to admit the existence of mental entities like beliefs, thoughts, desires, and feelings. As he liked to say, they are private events that go on inside a person's skin. However, he claimed, these events do not play a causal role in producing behavior. They are mere effects.

It is easy to see how such epistemological and methodological commitments can lead to the rejection of both Cartesian dualism and introspectionist theories about perceptions, thoughts, and feelings. Of course, we need to be clear that it wasn't just dogmatic allegiance to abstract epistemological principles that led Skinner to think that materialism, empiricism, and positivism were the way to go. Cartesian reliance on pure reason and psychologists' attempts to introspect carefully did not seem to be going anywhere. A new approach was needed.

But these commitments also mean that Skinner and his behaviorist colleagues were not very enamored of Freudianism, despite its materialist and supposedly scientific credentials. For even though the Freudians did not make any appeals to an immaterial soul and seemed to want to link their findings with neurophysiological results, their theoretical claims were highly suspect to Skinner. From the Skinnerian perspective, Freud's famous theoretical triad, the id, ego, and superego, were just that, theoretically postulated mental entities that were at best only loosely tied to actual environmental conditions and behavioral responses. In addition, as we shall see in more detail later, when the cognitive revolution in psychology began to blossom in the 1960s, Skinner's behaviorist commitments led him to be highly suspicious of the cognitivists' renewed scientific attempt to understand human behavior in terms of mentalistic categories. It is this same epistemological and methodological divide that has kept apart the behaviorists and the cognitive developmental psychologists, including Piaget and Kohlberg, who articulated moral developmental theories. Although we shall see later in this chapter that the current weight of scientific evidence is against Skinner's antimentalistic stance, we shall find that the behavioral explanations introduced by Skinner also have a place in a full account of human behavior.

So if mentalism was out for Skinner, we might think that neurophysiology might be in. But not so. Skinner was also committed to the autonomy of psychology as a discipline. This commitment was not based, of course, on a disbelief in neurophysiological processes or even a disbelief in the ultimate reducibility of psychological phenomena to neurophysiological phenomena. Skinner's reasons for claiming autonomy for psychology were more practical and pragmatic. First, there were the technical difficulties of finding and measuring neurophysiological processes that might be correlated with psychological variables and behaviors. Second, Skinner was not merely interested in a science of behavior for purposes of explanatory understanding. He was very much concerned with solving human problems. Consequently, he was interested in changing human behavior. Thus, like Francis Bacon, Skinner saw an intimate connection between science and technology. His science of operant behavior was a science precisely because it enhanced one's ability to predict and control. The ability to predict and control is essential for technological applications.

In Skinner's view, prediction and control were best accomplished with variables on the psychological, not the neurophysiological, level. So if one is interested in a technology of behavior, then one will turn to the psychological, and not the neurophysiological, level. We shall see in Chapter 7 that the autonomy of psychology thesis, a thesis also held by some who have taken the cognitive turn in psychology, has likewise been strongly challenged recently. This challenge comes both from those who believe in the ultimate reducibility of the mental to the neurophysiological, a position that Skinner did not oppose in principle, and from those who foresee the ultimate elimination of the mental. Skinner himself was an eliminativist of sorts. So, his adherence to the autonomy thesis does not put him in the radical conflict with those with reductionist or eliminativist proclivities as a similar adherence does some of his cognitivist colleagues. All this, though, anticipates discussions that we will have to pursue in more detail later. For now, we can just keep in mind that Skinner's account of human agency, including moral agency, does not put him into as much conflict with reductionist and eliminativist views of human agency as does a cognitivist account. So, even though he agrees with cognitivists about the autonomy of psychology from neurophysiology, he does so for different reasons.

I have briefly discussed Skinner's position with respect to various forms of mentalism and cognitivism in psychology and psychology's relationship to neurophysiology. There is one more piece of the big picture that we need to fill in before we can focus on the details of Skinner's view of moral agency. It might appear from our introductory paragraph that relationships between Skinner's science of behavior and evolutionary theory are completely cordial. But that appearance is indeed deceiving, for there is an aspect of the behaviorist tradition that is in strong opposition to the research traditions inspired by evolutionary theory of which ethology and, most recently, sociobiology are parts. We can summarize this opposition in two behaviorist theses (Erwin 1978). The first thesis states that most behavior of organisms, especially in humans and higher organisms, is learned rather than innate. The second thesis claims that there are general laws of learning applicable to all organisms. This latter thesis implies that learning is the same across species. Both of these theses have been drastically modified. We have already seen some good reasons for rejecting the first thesis since it seems to reflect an inadequate understanding of the relationships between evolution, development, and learning. As we shall see later in this chapter, striking experimental findings have led to the rejection of the second thesis. Nevertheless, I argue that the rejection of these theses does not lead to the total rejection of behaviorism. The laws of respondent and operant conditioning stand even if the simple dichotomy between genetically based and environmentally based behavior is no longer tenable and despite the fact that learning has species-specific characteristics. So even though our initial claim about the connection between a sociobiologically based account of moral agency and a behaviorally based account glossed over a multitude of necessary refinements, nevertheless it is, I think, a helpful oversimplification.

I have tried to situate Skinner's behaviorism both historically and with respect to other research traditions in psychology and relevant to psychology. But doing only this would leave a gaping hole in the big picture that I am trying to paint before we look at the details. As I mentioned earlier, Skinner is interested not only in finding out what makes us tick but also in improving our lives. For this reason Skinner, like his younger Harvard colleague E. O. Wilson, was concerned with much more than just what went on in his lab. He believed that a soundly based scientific psychology of behavior would have profound implications for individual humans and for society. Thus, throughout his career he was concerned about addressing the implications of behavioral psychology. His views on these implications are probably nowhere better expressed than in his utopian novel, *Walden Two* (1948), and his controversial presentation of behavioral psychology and its implications in *Beyond Freedom and Dignity* (1971). Like Wilson, Skinner rejects the separatist position between science and values and attempts to develop a scientifically based account of moral agency, one with profound social and political implications. To solve our major personal, social, and political problems, he insists that we have to give up our traditional religious and humanistic views of ourselves. Such views are formulated in terms of a mentalistic vocabulary. We think of ourselves as free agents acting on the basis of our beliefs about what is valuable and our desires to attain it. In other words, our ordinary view of ourselves, of who we are and why we do what we do, is formulated in folk psychological terms. But Skinner contends that such a view of ourselves is useless if we want to learn how to explain, predict, and control our behavior and its consequences. If we are to solve our pressing problems, Skinner is convinced that we must abandon our folk conceptions. To put Skinner's point in a paradoxical way, we can and ought to choose to move beyond freedom and dignity, and in so doing we will better ensure both. We are now ready to look at the details of Skinner's proposal that we develop a science of values.

I first examine Skinner's science of behavior and values.[1] Then I evaluate Skinner's claims and assess their adequacy. Although we shall discover that Skinner's project is not without severe problems, I argue that it provides things of value for my account of moral agency that I retain and build on. So, I conclude by considering how Skinner's project needs to be supplemented and put into a larger theoretical framework that attempts to deal with the sorts of mental causes that he refused to admit. Thus, just as I have used Skinner's behavioral psychological account of moral agency to supplement the purely biological science of values proposed by Wilson, I will use social cognitive accounts of learning and social cognitive theorists to construct a more adequate account of moral agency than Skinner's.

5.2   B. F. SKINNER'S SCIENCE OF BEHAVIOR AND VALUES

The science of behavior

B. F. Skinner has long been concerned with the many problems facing our world. In *Beyond Freedom and Dignity* and other works (e.g., 1953, 1976), he sketches these problems, some proposed solutions to them, and his own views about how to solve them using behavioral psychology. We know what these problems are: overpopulation, environmental degradation, poverty, injustice, war, the threat of nuclear weapons and technologies. If anything, the problems that Skinner details have gotten worse since he wrote his book in 1971. In order to solve these problems, Skinner believes that we need to substitute a scientific view of humans, based in part on his own behavioral psychology, for the commonsense, religious, and humanistic views we have of ourselves. Although these latter views differ in various ways, they all share the conception of humans as doing what they do because they have certain wants that they desire to satisfy and certain beliefs about how these wants can best be fulfilled. According to this conception, we choose between our various wants on the basis of the values we assign to them and on the basis of what is demanded of us by our moral obligations and values. People may differ about what the ultimate basis for moral values and obligations is. Religious people may say that they come from God; others, like some humanists, may claim that they come from ourselves and our creative choices. Still others may maintain that our obligations and values come from society. But whatever their source, all agree that we choose freely to satisfy or not to satisfy certain wants on the basis of our moral values and obligations.

Skinner, however, has different ideas. Extrapolating from his study of animal behavior and human learning, he believes that we must substitute a causal explanation of why and how we do what we do for the commonly accepted account based on free choice. Skinner and other behavioral scientists maintain that our behavior, like that of other animals, is governed by identifiable causal factors. These causal factors and the effects that they bring about can be formulated in some basic laws. The two basic laws of behavior are those of respondent and operant conditioning.

The Russian physiologist Ivan Pavlov is responsible for the earliest formulation of the law of respondent conditioning. Pavlov knew that a dog's mouth watered whenever it was presented with a piece of meat. He decided to see if other things, stimuli, would do the same thing. So he rang a bell before he presented his dog with some meat. Ordinarily, dogs do not salivate when they hear a bell ring. Salivation is the result of something like meat. After Pavlov had repeated his trials for a while, it turned out that his dog began to salivate when he heard the bell ring and before the meat was presented. The meat, the unconditioned or unlearned stimulus, elicited or brought about salivation, an unconditioned or unlearned response. The bell, the conditioned or learned stimulus, elicited the salivation response also. The dog had learned to salivate at the ringing of the bell. It turns out that there is a whole set of unconditioned stimuli, different for different animals,

106

that brings about different unconditioned responses. And these unconditioned stimuli can be associated with other, nonrelated stimuli, like the bell with the meat, to bring about the same behavioral effect as the unconditional stimulus. In fact, a whole chain of learned stimuli can be associated with the unconditioned stimulus and, thus, evoke the same behavior that the unconditioned stimulus does. This kind of learning is called *respondent conditioning* or *respondent learning,* and the elicited behaviors are called *respondent behaviors.*

What about the laws of operant behavior? Suppose that you saw a dog roll over and play dead, that is, stay on his back, motionless, with his legs extended stiffly in the air. The first thing you would probably notice is that his behavior is rather unusual unless, of course, he really is dead. Dogs salivate when presented with meat without learning to do so. But they do not roll over and play dead naturally. So the behavior is not one that could have resulted from respondent conditioning since in respondent learning unconditioned stimuli evoke naturally occurring behaviors. How does a dog learn this new behavior? A dog sometimes rolls over when he is playing or when he meets a dog that is his superior. A trainer, then, can be alert to when this happens and reward her dog when he rolls over. Gradually, by the skillful application of rewards, she can get her dog to roll over and keep his body stiff and his legs straight up in the air. Here's an example of another type of operant learning. Suppose you're teaching your dog to walk on the leash. At the beginning, he tugs hard and almost drags you along. But you hold firm and walk at a moderate pace. After a while he, too, is walking along at your pace. This time, instead of rewarding him for doing something, you have presented him with a negative consequence for the continued behavior of tugging at the leash: the collar digging into his neck in an uncomfortable fashion. So, he learns to avoid that consequence by walking at the right pace. One more example. Suppose you're in the park walking your dog, and you decide that it would be good to let him off the leash so that he can have a good run. But you know that when you did this before, he tended to run off and you had a hard time finding him. When this happened in the past, you had to hunt for him and then chase after him to get him back on the leash. Sometimes you gave him a little whack on his rear end when you finally caught up with him, hoping that this punishment would decrease his running-off behavior. You're not sure that it has done so. Sometimes he may avoid you because of the whacks!

Thus there are two kinds of operant learning, reinforcement and punishment. The first two cases were examples of two kinds of reinforcement, positive and negative. The last case was an example of punishment.[2] The chief difference between reinforcement and punishment is that reinforcement increases the chances that the behavior that precedes the reinforcer will occur. In positive reinforcement the rewards, the positive reinforcers, increase the probability of the behavior for which they are a reward. The second type of reinforcement learning is negative reinforcement. Negative reinforcers, like positive ones, increase the probability of a behavior. But negative reinforcers increase the probability that the behaviors for which they are the consequences will be avoided. In punishment, the consequence of the behavior diminishes the chance of that behavior occurring rather than increasing it. One of the things that Skinner and his colleagues discovered was that reinforcement learning, particularly positive reinforcement, is a more powerful mode of learning than punishment.

Although our examples concern animal behavior, Skinner holds that the same principles apply to human learning. For example, students might condition their teacher to stand in one small section of the classroom near the blackboard by reinforcing him with smiles and nods of approval when he is in certain areas of the room. At the beginning, he roams the classroom while lecturing and discussing with his students, but by a judicious use of reinforcers, the students can confine their teacher to a corner of the classroom. We can explain why the teacher teaches from the "right spot" in the classroom by means of the laws of operant conditioning.

At this point, let's pause for a moment and consider what kind of an explanation is provided by operant conditioning. One clear hint of an answer to this question is that operant explanations are presented in terms of consequences rather than antecedents. We recall that explanations of adaptations also focus on consequences rather than antecedents. Both evolutionary theory and operant theory are types of selection theory. As with an adaptation, we answer the questions of how the adaptation occurs and why it occurs. Of course, we do not answer all the how and why questions that arise in investigating the phenomenon. For instance, we might want to know what component behavioral capacities are associated with the molar (i.e., large-scale) capacity of teaching in the right spot and how they work together to bring about that sort of behavior. Or we might want to know what other, if any, functions are served by teaching in the right spot. But it should be clear that in this case the answer to the question of why the teacher is teaching in the right spot is presented in terms of its effects, namely, that it has in the past increased the chances of receiving student approval. This answer provides an explanation in terms of the past consequences of the behavior in question. Alternatively, to answer the question of how the distribution of teaching locations has occurred is to appeal to key characteristics of the teacher in question and to the critical features of the students who constituted the selecting environment, that is, the various behavioral capacities of the teacher, including being able to teach in the right spot, and the visual discriminatory capacities and selectively reinforcing abilities of the students. We explain how the results of the interaction occurred in terms of these key antecedent characteristics of the students and the teacher.

So, both adaptation explanations in evolutionary theory and operant explanations in behavioral theory are consequentialist explanations. I noted earlier, when discussing consequentialist explanations, that they have been called teleonomic explanations to distinguish them from teleological explanations, especially purposive explanations. The reason that this distinction has been made is to make it clear that these sorts of explanations do not involve thoughts, desires, and purposes. Just as Darwin did not substitute the choices of nature for those of God, so too it is Skinner's goal to explain end-directed activity without appealing to choices. Whether he is successful in doing this is a question that we shall be examining shortly. Indeed, you have probably noticed that even the teacher's behavior of teaching in the right spot need not be explained in terms of the teacher's decision. He did not decide, after careful consideration of all the alternatives and their pros and cons, that spending most of his time in the right spot is the most educationally effective choice that he could make. Nevertheless, the actions of the critical environment, the students' actions, seem to be loaded with mental processes. But setting that issue aside for the moment, it should be clear that the operant explanation did not invoke any mental activity on the part of the teacher. It is as mindless as natural selection. And if Skinner is right, the students' selective activity, which in our example seems very much like the artificial selection of breeders, can itself be redescribed and explained without mentalistic baggage.

Given the similarity between explanations of adaptations and operant explanations, the question naturally arises about their relationships. As I suggested at the beginning of this chapter, there seems to be a natural connection between the two sorts of explanations, one that both sociobiologists and behaviorists, seeking to give an elegant and simple account of moral agency, might be delighted to find and exploit. Remember that what we are doing is trying to illustrate the laws of behavior. According to Skinner, these basic laws govern what we do. Of course, our behaviors are much more complicated than standing in a particular place. But the idea is the same. According to Skinner, our everyday activities, even though much more complicated, can be explained by these basic laws. Naturally, the explanations would be much more intricate because there are many more stimuli, reinforcers, and punishers involved in complicated causal chains bringing about many different kinds of behaviors. Skinner's claim is that we can analyze the stimuli, behaviors, reinforcers, and punishers involved in the problem-causing behaviors that we are all now engaged in, and then we can figure out how to change the stimuli, reinforcers, and punishers to produce different behaviors, ones that do not cause the difficulties that seem to be overwhelming us. We can learn behaviors that lead to positive results for ourselves and everyone.

Where do these long chains begin? A natural answer to this question is to say that they must begin with reinforcers that are not learned but are rather built in genetically. Nature provides us with a set of basic behaviors that are reinforced in normal environments and that promote fitness. It is not hard to imagine what some of these are: food, water, warmth, protection, care, companionship, and sex. The behaviors that bring about the acquisition of these items are increased in contexts with the proper stimuli. We can give a similar analysis for negatively reinforcing events and for punishment. So the science of operant behavior seems to have a clear link with sociobiological explanations of behavior. The next step seems obvious, and it brings us to Skinner's next major thesis.

The science of values

For the time being, let's grant that Skinner has come up with laws that govern what we do. The question of why is answered by saying that we do what we do because it is reinforcing. But this hardly seems satisfactory. If we are going to solve our problems, we need to know the goals for which we act, and we need to know the right kinds of goals, correct values, not just that the consequences of what we do increase the probability that we will do the same thing. So, we must discuss Skinner's answer to the why question in more detail. We need, in fact, to address directly his provocative claim that the science of operant conditioning is the science of values.

To do this, I focus on the notion of positive reinforcers. In Skinner's view, values are positive reinforcers and disvalues are negative reinforcers. Since the behavioral scientist studies the ways that reinforcers govern our behaviors, she is necessarily concerned with the way values direct our behaviors. Insofar as she can discover a systematic account of the relationships between positive and negative reinforcers and our actions, she is developing a science of operant conditioning that is, by the nature of positive and negative reinforcers, also a science of values.

How is it that we can describe positive reinforcers as values? In Skinner's view, it would be a mistake to confuse reinforcers and punishers with the pleasures and pains that, from a commonsense point of view, we often identify with values and disvalues. Indeed, Skinner is neither a hedonist nor a utilitarian. Recall that, technically, positive reinforcement is the effect or consequence of a behavior that increases the chances that the behavior will be performed again. Pleasures are sometimes, but not always, a result of positive reinforcers, and pains are sometimes the results of punishers. And in avoiding negative reinforcers we, on occasion, avoid painful events. But pleasures and pains are experiences that we sometimes have as a consequence of positive reinforcers and punishers. They are inside us; the positive and negative reinforcers are not. Values are those things, properties of things or events that, because they are positively reinforcing, increase the chances of a behavior happening again.

But what are the kinds of things or activities that induce us to continue to seek them? Putting the question that way helps us to see what Skinner has in mind. Food, clothing, shelter, the things that make for good health, friends, family, learning, and athletic and artistic activities are some positive reinforcers. Notice that some positive reinforcers are so only for individuals. For instance, you may be reinforced by spinach. I may not be. Other reinforcers are cultural and social. North Americans are more reinforced by football than by soccer; for South Americans, the reverse. So, there are individual, cultural, and societal differences in reinforcers and, thus, in values. But Skinner also claims that there are values that are common to all of us as members of a species. These specieswide values are, therefore, cross-cultural and cross-temporal. They can be found in all cultures and in all times. All of the things on our list of examples would count as specieswide values. Of course, there are individual, cultural, and societal variations on the kinds of food, clothing, shelter, and so on that people find reinforcing and so valuable. That is why it is important, in applying the science of values, for the technologists of behavior to know these individual, cultural, and societal or group variations.

We realize that identifying specieswide values gives us only a set of general laws about which things or events usually increase the chances that humans will perform those behaviors that enable them to achieve those things and realize those events. To have a science, we need not only general laws to predict and explain individual phenomena, but also theories to explain the general laws by enabling us to understand the basic realities behind the phenomena captured by the laws. So, if Skinner is serious about offering us a science of values, he needs to answer the question of why positive reinforcers are reinforcing or why values are valuable.

At this juncture, behavioral psychology connects with evolutionary theory. What do adaptations and fitness have to do with the science of operant conditioning and the science of values? The evolutionary answer to that question is that some reinforcers are things that help the reproduction and survival of organisms. That is why they are reinforcing and valuable. Better-adapted organisms are more fit than less well adapted ones because they achieve these reinforcers more successfully. As we have seen, the key assumption here is the integrationist's sociobiological one that survival and reproduction are valuable. If we make that assumption, then we can understand why many of the reinforcers on our list, especially what we called the specieswide reinforcers, are reinforcing. It is clear that food, shelter, clothing, and family contribute to reproduction and survival. So, the former are means to the latter. But there is another way to understand how evolutionary theory helps explain why reinforcers are reinforcing: to notice that reproduction and survival are necessary for the achievement of other reinforcers. From the Skinnerian perspective, we don't have to say that survival and reproduction are the only ultimate reinforcers; there can be others. But without survival and reproduction, most, if not all, of the other reinforcers are unachievable. If we don't survive, we cannot expect to be positively reinforced in other ways; nor will our children if we don't reproduce. The same thing is true if we now consider positive reinforcers as values. Survival and reproduction, as one set of ultimate values, help explain why other values identified by the science of operant behavior are valuable. They are valuable because they enable survival and reproduction. And the latter are valuable as necessary conditions for the achievement of the values discovered by the science of operant behavior. So Skinner maintains that the science of operant behavior or, more broadly, behavioral science fits in with biology, in particular with evolutionary theory, and that the two together can offer an explanatory and predictive account of values. Having examined Skinner's account of moral agency and its connection with a sociobiological account, we are now ready to assess it.

5.3 ASSESSING SKINNER'S BEHAVIORAL ACCOUNT OF MORAL AGENCY

Skinner and the questions of moral agency

At first glance, Skinner's account of moral agency seems so far from the four-tier model of moral agency presented earlier that it may appear to us that little or nothing of it can be salvaged and that, if a biologically based account of moral agency is going to make satisfactory links with psychology, it will have to look elsewhere. Recall that we have said that any satisfactory account of moral agency will have to provide credible answers to the key questions of relevance, acquisition, action, and justification. Take, for instance, the question of justification, which we shall discuss in some detail in Chapter 8. Justification concerns, at a minimum, whether an account of moral agency can provide suitable justifications for moral beliefs, motivations, and actions. Offering such justifications is usually understood to be a matter of providing good reasons. But if Skinner's account of agency, including moral agency, excludes any role for mental factors such as reasons for an action, then it seems to be fundamentally incapable of even beginning to address the question of adequacy. Even setting aside the question of justification, Skinner's account seems to be in deep trouble. His view of agency seems to make his account of moral agency both functionally irrelevant and unable to explain how humans attain their moral capacities and put them into action. His position on the nature of values as reinforcers appears to render his account of moral agency substantively irrelevant or, at best, marginally relevant. Thus Skinner's account of moral agency, on the surface at least, seems unable to answer satisfactorily the questions of justification, relevance, acquisition, and action.

I believe that these are serious charges against Skinner's views and that they have to be considered very carefully. However, I do not believe that they force us to reject Skinner's proposal in its entirety. Rather, it can be modified in some important ways to take into account both what is necessary for moral agency and what further developments in cognitive psychology show us to be necessary for an adequate account of human agency. First, I consider whether Skinner's account of the nature of values renders his view substantively irrelevant.[3] Then I address questions of acquisition and action. Although the criticisms concerning relevance and the replies to these criticisms may appear to leave open the question of the relevance of Skinner's views for moral agency, the final criticisms concerning Skinner's rejection of cognitive capacities will prove decisive in forcing us to move beyond his behaviorist position and to embrace an account of both human agency and moral agency that includes cognitive abilities.

Are Skinnerian values empty?

Even though Skinner includes such apparently nonmaterial values as family, friendship, and artistic and intellectual pursuits on his list, and maintains that since they function as positive reinforcers they are legitimately there, the critic may feel that they have no real home in the Skinnerian kingdom of values. They are ill-fitting aliens whose home is really in commonsense, humanistic, and religious accounts of values. Aristotle provides the critic with one way of putting this feeling of uneasiness into an objection. Aristotle maintains that values are built on human nature. By that he means that we have a set of capacities that constitutes our nature and whose development and exercise make us well-functioning human beings. Besides our bodily needs for food, clothing, and so on, we have capacities for knowledge, love, friendship, and other nonmaterial goods. Thus, what makes something valuable is that it promotes good functioning by fulfilling the capacities of human nature. But since for Skinner values are only reinforcers, he can identify a value only as something that makes someone increase his or her activity in its pursuit.

In order to answer this objection, the Skinnerian asks us to recall the connection between her behavioral science and evolutionary theory. Operant conditioning depends on genetic capacities. Learned behaviors and secondary reinforcers are founded on unlearned behaviors and primary reinforcers. What this means concretely is that our chances of survival and reproduction have been bettered because we have certain unlearned capacities and capabilities, such as our cognitive powers and our abilities to cooperate and to form social groups. These abilities and capacities have themselves been enhanced and extended to a wide range of behaviors through learning. We can include in these learned achievements such things as culture, ethics, science, and religion. So there is a way that Skinner can make use of the Aristotelian notion of human nature without all the Aristotelian baggage about species being eternal. Human nature is a historical, not an ahistorical, eternal phenomenon. On that score Skinner and Aristotle differ. But they can agree in large measure about the needs, capacities, and capabilities that make up human nature. So Skinner, as well as Aristotle, can appeal to human nature to explain why values are valuable.

But the critic, still unsatisfied, objects that Skinner's concept of human nature is biologically reductionistic. If something is valuable only because it promotes survival and reproduction, then only two things are valuable in themselves; the rest are valuable only instrumentally, that is, because they lead to what is intrinsically valuable. Recall, however, that Skinner need claim only that survival and reproduction are necessary conditions for the continuation of such values as culture, ethics, science, and religion. Thus the former are means to the latter just as much as the latter might be means to the former. Moreover, unlike biological reductionists, Skinner does not claim that culture, ethics, science, and religion must somehow be biological adaptations, that is, that they are the result of natural selection acting on heritable variations. Most probably they are not. Rather, they are, as most psychologists and others would claim, learned, not inherited, traits. Thus, even though they might contribute, on occasion or frequently, to survival and reproduction, they could, from Skinner's perspective, have an independent value.

But this response doesn't really get Skinner out of the woods, since to claim that these values are learned and possibly independent of the biological basis of survival and reproduction seems to commit Skinner to another foundation for values and to certain values that are nonmaterial, involving the realm of the mind. And, of course, Skinner has made his reputation, in part at least, by his denial of the existence of the mind and any causal role for beliefs and desires in effecting our actions.

Thus the fundamental objection remains that Skinner views humans as empty organisms. They lack a subjective side, and since that is essential for having values, Skinner's account of values is ultimately hollow. In order to present the Skinnerian reply, we need to introduce another important concept in the science of behavior: the notion of *learning history*. Consider Lisa and Al. They have been going to church and church school for quite a while. And they have been reinforced and punished for their behaviors for a substantial period of time. Indeed, they are considered to be good, even though not perfect, young adults. Their past experiences constitute their learning history. Because of their learning history, Al and Lisa tend to do certain things and not others in a given situation. They tend to do what is right when it comes to the environment or helping someone out who is in need. Does Skinner, then, deny the existence of thoughts and so the existence of values? In one sense he does. He does not believe that thoughts as we usually think about them exist. According to the commonsense view, they are things in our minds that enable us to know about the reality outside of our minds and to act on it. In Skinner's view, our ideas are the consequences of our actions and the actions of the reinforcers. So, they are, in part, another way to talk about learning history. But they are not the primary movers of either our learning achievements or our actions. The natural and social environments and their reinforcers are the primary movers. We do not act or change because we are moved by our ideas or have changed them. Our ideas change because we have acted or acted differently. Every individual has a genetic history and a learning history. No individual is a blank slate or an empty organism. But our genetic and learning histories, as well as our actions, come from somewhere; they are shaped by the natural and social environments that produced our evolutionary history as a species, our cultural and social history as cultural and social beings, and our individual learning histories as unique persons.

But when we talk with Al and Lisa, they don't tell us that they clean up the river bank because of their learning history and their current schedule of reinforcement. They tell us that they believe that the preservation of the environment is an important value. Moreover, they indicate that they think it is their duty to maintain it. They desire to do so and think the project of cleaning up the river bank is an important way to fulfill their desire and do their duty. In addition, they tell us that these beliefs and desires and the actions that result from them are themselves based on their religious beliefs, duties, and the desires to fulfill the latter. Our Skinnerian will, of course, allow this kind of talk, but she will also remind us that it is part of a folk psychological account of why we do what we do. She will urge us to be careful not to confuse common sense with science. Commonsense physics tells us that if something moves, it needs a cause, a force, to move it. But Newtonian physics reminds us that if a body moves at constant velocity, it doesn't need any force to keep it in motion. So, our Skinnerian will counsel our friends to feel free to talk about beliefs and desires and say that they are the causes of actions, but to realize that such talk has to be replaced by discussions of discriminative stimuli, learning histories, and reinforcement schedules. Folk psychology is fine if it recognizes its place and stays there. After all, we still talk about sunrises and sunsets and use Earth-centered astronomy to navigate. Folk psychology has practical advantages, but it cannot take the place of scientific psychology. And scientific psychology for the Skinnerian is, of course, Skinnerian behaviorism.

How adequate is Skinnerian behaviorism?

From the 1920s to the early 1960s, psychology was dominated by various forms of behaviorism that in one way or another denied a causal role to cognitions in human behavior (Bower and Hilgard 1981). So, to oversimplify a very complex story, until the early 1960s it was probably reasonable for Al and Lisa to accept the Skinnerian's advice. But psychology has gone through a cognitive revolution in the last 30 years or so. In fact, two cognitive revolutions have recently occurred in psychology, the first beginning around 1960.[4] The behaviorist hold began to loosen, first in areas that had come to be called cognitive psychology, like those of perception, imagery, thought, memory, problem solving, and language (Gardner 1985). Somewhat later, those areas of psychology concerned with agency, such as learning, personality theory, and social psychology generally, began to be influenced by the winds of cognitivism. To put it too succinctly, the first revolution made cognition an acceptable part of the input side of the equation for human agency, and the second did the same for the output side. Theories of agency and moral agency have to do primarily with the output, not the input, side of the equation. In the most recent edition of their magisterial account of learning theories, Bower and Hilgard (1981) have argued that cognitive social learning theories, as theories of agency, have taken up the gains in cognitive psychology and begun to apply them to problems of agency. They contend that such cognitive social learning theories as Albert Bandura's, whose views we examine in Chapter 6, will serve as integrating theories that explain cognitively based human behavior while drawing on the developing accounts of cognition in cognitive psychology that originated in information processing theories and computer science.

We have already seen that Skinner himself was not opposed to the existence of cognitions. His stricture was that they be considered effects, not causes, of our behaviors. Thus, Skinner denied that we are empty organisms. He realized that we came into this world with an evolutionary history built in and that as we developed, we acquired a learning history. Both of these internal factors affect what we do, how we do it, and why. Behaviorists, in general, were not unwilling to discuss mediating variables in their explanations of human and animal behaviors. But, as we have seen, for epistemological and metaphysical reasons, as well as explanatory and methodological reasons, they claimed that even if the organism has a subjective side, it is still cognitively empty. The cognitive revolution in psychology and the rejection of positivistic views about science changed all this. If this change in accounting for agency is well established – and we shall see that there are good reasons for thinking that it is – then it will have an important effect on our account of moral agency. Because of their neglect of cognition, both the Skinnerian and the Skinnerian-cum-sociobiological accounts of moral agency turn out to be incomplete. Neither separately nor together are they sufficient for an adequate understanding of moral agency. Nevertheless, the Skinnerian-cum-sociobiological account of moral agency will find a place in a more comprehensive account of moral agency, one that tends to confirm our folk psychological view of moral agency as requiring cognitions as well as learning history and natural endowment.

So, let's see why Lisa and Al are now well advised, for scientific as well as pragmatic reasons, to hold on to talk about cognitions when trying to explain what they are doing. To do this, I examine the scientific reasons why agency, and consequently moral agency, is best accounted for by bringing in cognitions.[5]

Some empirical inadequacies of the behaviorist laws of learning

I have said that Skinner's behaviorism is based on two theories of learning: respondent and operant conditioning. The theory of respondent learning tells us that whenever a neutral stimulus is paired with one that evokes an unlearned or unconditioned response, that neutral stimulus comes to evoke the same response. The theory of operant learning tells us that all reinforcers increase the probability of a response and all punishers decrease the probability of a response. What empirical support do these theories have? More precisely, what sort of empirical support is there for the theory that human agents acquire behavioral patterns and put them into action on the basis of these learning principles? Let's focus on operant learning.

We should note that the question presupposes something that some critics of behaviorism, specifically critics of the law of operant conditioning, have not been willing to admit: that the law of operant learning is, indeed, an empirical law. It is easy to see why they might be suspicious of its empirical standing. Take the definition of a reinforcer as something that increases the probability of a response and a punisher as something that decreases the probability of a response. Now consider the law of operant learning stated earlier and substitute in our definition of a reinforcer. It becomes: Everything that increases the probability of a response increases the probability of a response, and everything that decreases the probability of a response decreases the probability of a response. Behold, we have another example of the indefinitely many eternal truths that inhabit the nonempirical realm! Using the earlier definitions of reinforcers and punishers, our supposedly empirical laws have been turned into tautologies whose truth or falsity is completely independent of empirical considerations.

Of course, what one might do in the hope of remedying this situation is to identify certain things, events, or states of affairs as the sorts of entities that, for instance, increase the probability of a response in specified circumstances. That's what we did earlier, listing such things as food, clothing, shelter, and the like as reinforcers. Finding a list of primary reinforcers, we can then claim that other stimuli can increase the probability of a response by becoming associated with these primary reinforcers. We could do the same thing with punishers. We now have empirical claims, claims that can be tested. If we are cautious – and well we ought to be – we might want to qualify our laws by throwing in a ceteris paribus clause, that is, a "the rest of the relevant factors being the same" clause. That is, we might want to say that x, y, and z increase the probability of a response in a given circumstance if everything else has stayed the same. We might then go on to assert that that's usually the case. So our laws usually hold. An obvious problem with this cautious strategy is that it makes our laws too vague. We always have a loophole in case we run into circumstances in which the law seems to fail. We can just claim that those circumstances in which the law failed to hold are ones where everything else didn't stay the same. But this kind of escape from falsification will get us nowhere in the long run. So, to avoid the vagueness and the untoward consequences of ceteris paribus clauses, we have to strengthen our formulations. How strong do we need to make these claims?

In a classic article, Paul Meehl (1950, as cited in Erwin 1978) not only suggested a way to avoid the charge that the law of operant learning – the *law of effect,* as it is often called – is a tautology, but also offered some formulations of it that have empirical bite. He proposed a weak and a strong version of the law of effect. The strong version states that every increase in the strength of a response is due to a *transsituational reinforcer.* Strengthening a response increases its probability of occurrence. By a transsituational reinforcer, Meehl meant a reinforcer that strengthens all learnable responses. A learnable response, according to Meehl, is a response that can be strengthened by exposure to a situation rather than by surgically induced, drug-induced, or maturational changes. The weak version of the law of effect states that all reinforcers are transsituational. This comes down to the claim that a reinforcer can be used to increase the probability of any learned response (Erwin 1978). It doesn't add, as does the strong version, that any increase in the probability of a response is due to a transsituational reinforcer. How satisfactory are these versions of the law of effect?

When we think about it, we can see that the weak version of the law is too weak. It could be true that a reinforcer can strengthen the probability of any response that can be learned, and yet still be the case that most learned human behavior is not governed by reinforcers. Reinforcers may be sufficient for the increase in the probability of a behavior but not necessary. On the other hand, the weak law may also be too strong. If reinforcers are transsituational, then we would expect that the same reinforcer must be reinforcing in any situation. But that does not seem plausible. A compliment may reinforce a conciliatory attitude on the occasion of a minor dispute but do nothing when the dispute is one of principle. Reinforcers and punishers seem to be relative to species, as the famous experiments of the Brelands with pigs and those of Garcia with conditioned taste aversion in rats have shown (Breland and Breland 1961, 1966; Garcia and Koelling 1966; Revesky and Garcia 1970, all cited in Bower and Hilgard 1981). In addition, David Premack has argued that reinforcers are relative to each other (Premack 1959, 1965, 1971, all cited in Erwin 1978). If, as seems to be the case, people have different preferences for various things, states of affairs, and activities, then reinforcers are not transsituational. For instance, if you prefer doing A to B, B to C, and C to D, then you can increase the probability of your doing D by reinforcing it with either A, B, or C. But you cannot increase the probability of doing B, for instance, by following it up with C. Roughly, this is what has come to be called *Premack's principle.*

On the other hand, the strong law of effect seems not to be supported when applied to humans. Most of the data used to support the law come from studies of rats, pigeons, and other nonhuman animals in restricted experimental situations (Estes 1971, as cited in Erwin 1978). And if, as seems to be the case, reinforcers are not the same for all species, the strong law does not even hold for all nonhuman species.

But the case against the sufficiency of the laws of learning in accounting for human behavior is stronger than just the claim that the evidence is not sufficient to establish those laws in the case of human learning. Rather, there seems to be positive evidence that in humans, at least, in some cases, if not in many, they are false. First, there seems to be evidence that both respondent and operant conditioning depend on awareness of pairings and contingencies of reinforcement. In addition, imagery, beliefs, and expectancies seem to be necessary to account for human learning.[6] Brewer (1974), in a well-known and widely discussed article, argues that there are clear experimental designs, called *disassociative designs,* that allow one to distinguish between conditioning and cognitive explanations. In these studies, the learning or performance of subjects who are unaware of or uninformed about the contingencies of reinforcement are compared with those of subjects who are aware of or informed about them. The latter group of subjects display superior learning or performance. Brewer has two conclusions. His more modest conclusion is that explanations of learning in humans ordinarily require an appeal to cognitive factors, although there may be some conditioning of autonomic responses in humans. His more radical conclusion is that there is no evidence for pure conditioning explanations in the learned behavior of adult humans. Learning theorists like Dulaney have attempted to respond to Brewer's claims by invoking auxiliary hypotheses to explain away the results of the cases discussed by Brewer (Dulaney 1974, cited in Erwin 1978). Erwin (1978), to whose careful account of the fortunes of the laws of learning I am much indebted, concludes cautiously that the appeal to auxiliaries prevents the conclusion that the experimental work cited and summarized by Brewer is sufficient to refute the claim that the laws of operant and respondent conditioning can account for human learning or significant portions of it. Nevertheless, even though the appeal to auxiliary hypotheses may lead one to the conclusion that conditioning accounts of human learning are just as good as cognitive accounts, they do not lead to the conclusion that they are superior to cognitive accounts. His final assessment is that "[in] the absence of other ways of proving the superiority of the conditioning explanation, and assuming that other types of conditioning experiments are inadequate, then the conditioning literature fails to provide any firm support for the idea that a behavioristic theory of learning will eventually prove to be adequate" (Erwin 1978, p. 112). Indeed, Erwin believes that the results of the disassociative design experiments do tell against the truth of the claim that the laws of learning are sufficient to account for a significant portion of human learning.

Moreover, there is evidence that some behaviors of nonhuman species involve cognitive factors. In a famous set of studies, Martin Seligman and his colleagues (Seligman and Maier 1967; Seligman 1975; Maier and Seligman 1976) identified a phenomenon that they called *learned helplessness.* One experiment was conducted in two phases with different groups of dogs. In the first phase, dogs were restrained in a hammock and given many unpredictable painful shocks. The dogs in this group could escape by pushing a panel. Other dogs received the same shocks as those in the first group, but they could not do anything to escape or control the shocks. The first group of dogs apparently learned that the shock was controllable. The second group learned that it was not. In the second phase, each dog was trained in how to avoid shock using a two-way shuttle box; in response to a conditioned stimulus, a dog learned to avoid a shock by immediately jumping over a barrier separating the two chambers. The response was always followed by a termination of the conditioned stimulus and avoidance of the impending shock. In this situation, the dogs from the first phase of the study who had been able to escape the shocks quickly learned how to escape in this new situation. So did a control group of dogs who had not been shocked in the first phase of the study. However, the animals in the second group who had received inescapable shocks in the first phase were practically helpless during the second phase. They did not often jump to the other side to avoid the shock. Rather, they remained still, accepted the shock, and whined. They did not learn to escape. Their inability to escape was aptly termed learned helplessness.

These sorts of experiments have been replicated with other species, including humans. Moreover, learned helplessness seems to occur with positive reinforcers as well as negative ones. In addition, generalization seems to occur: If helplessness is learned in one situation with one sort of reinforcer, it transfers to a different situation with different reinforcers. Animals seem to learn not only that the responses they have tried will be ineffective, but that all responses will be ineffective and that the shocks are completely independent of their responses. Moreover, they seem to generalize to other situations and reinforcers. Behaviorists have attempted to explain learned helplessness in terms of superstitious behavior. For instance, the animal's sitting, frozen position may have been reinforced because that just happened to be the animal's response at the time that the shock was terminated. So, the animal learned an incorrect association. This learned response could then compete with learning the correct escape response.

Maier and Seligman reject this alternative hypothesis and other alternative explanations of the phenomenon (Maier and Seligman 1976).[7] Given the robustness of these findings, it seems there is evidence that not only learning theory accounts of human behavior are false, but also that some learning theory accounts of animal learning are false. These need to be supplemented by cognitive accounts. In the case of learned helplessness, what seems to be required is an appeal to an animal's cognitions about the relationships between its responses and the reinforcers present in the situation.

Where does all this leave the Skinnerian? The evidence we have summarized is telling against claims for the sufficiency of behavioristic learning theory and is indeed indicative, even if not definitively so, of its falsity. At this point, we should recall that scientific cases are not built on empirical evidence alone. Conceptual connections with other parts of a scientific discipline and with other scientific disciplines, as well as methodological and philosophical considerations, may play a part in the assessment of a scientific position. The Skinnerian and the cognitivist share some metaphysical convictions about materialism and methodological convictions about the importance of empiricism in the broad sense of linking as much as possible one's theoretical and empirical claims. We have seen that the Skinnerian cannot rule out the mental merely on a priori methodological commitments that limit the role of the mental to the status of a dependent variable. I have also argued that the major metaphysical difference between Skinnerians and cognitivists on the role of mental factors in our lives is one that has to face the court of empirical evidence. At this point, at least, we have some reason to believe that Skinnerians have failed to make their case and that there is some positive evidence for the cognitivist account.

5.4 BEHAVIORAL CAPACITIES AND MORAL AGENCY

Where does all this leave us? Recall that the critic had challenged the Skinnerian view because it makes a mockery of moral agency. I have examined these objections and discussed ways in which the Skinnerian can attempt to dissociate herself from the charges. But in the end, I have concluded that the cognitive revolution in psychology has shown that the Skinnerian view of how we acquire our capacities as agents and put them to work fails because it does not take into account our cognitive capacities. Because of this failure, I have indicated that we will have to revise and expand the Skinnerian account of agency. So, even though we might concede that the Skinnerian account of moral agency is substantively morally relevant, nevertheless, because it does not adequately answer the questions of acquisition and action, it cannot serve as the complete scientific psychological basis for moral agency.

Consequently, our initial idea of putting the results of sociobiology and behavioristic psychology together to give us a complete account of the biological and psychological bases of moral agency, at least with respect to how we acquire moral agency and put it to work, fails. Neither the sociobiology of Wilson nor Skinner's science of operant conditioning, alone or connected, is enough, not merely because they do not display the four-level model of moral agency that we have suggested but, more important, because they are not scientifically adequate. This, of course, does not mean that we need to abandon either sociobiology or behaviorism. Rather, as I indicated earlier, my account of moral agency requires that we include explanations of the role of the biological and noncognitive components of our complex capacity to be agents and moral agents. It is here that sociobiology and behaviorism can make a contribution. As our brief look at the cognitive critique of Skinnerian behaviorism has indicated, these explanations have to fit into a more adequate account of the role of our cognitive capacities in the acquisition and exercise of moral agency. It is to that account that we are now ready to turn.

NOTES

1. I omit a discussion and evaluation of the third major part of Skinner's proposal, a technology of values based on his theory of behaviors and values (cf. Rottschaefer 1995).
2. There are actually two types of punishment learning, but we don't need to discuss them for our purposes.
3. I shall not address directly the charge that Skinner's account of moral agency is functionally irrelevant because it denies human freedom. I have argued elsewhere (Rottschaefer, 1995) that Skinner's thesis that the science of behavior is the science of values leads to a Platonic characterization of freedom as the ability to do the good where the latter is characterized in terms of positive reinforcers. Thus Aristotelian-based critiques that Skinner denies freedom of choice beg fundamental questions. My critique of Skinner focuses on the scientific inadequacy of his answers to questions of acquisition and action. By incorporating Skinnerian learning mechanisms as a first-level component in my model of moral agency, a model that enables a soft-determinist account of freedom, I attempt to sidestep charges of the functional irrelevance of the Skinnerian account of agency.
4. I do not want to make too much of the notion of revolution; its use has become something of a cliché. In this context, I intend it in the relatively neutral sense of an important shift in theoretical and empirical scientific work, not in its Kuhnian sense of a relatively discontinuous transition from one period of normal science

123

to another in a particular scientific discipline. For discussions of the nature of the cognitive revolution in behavioral psychology, see Dobson (1988), Erwin (1978), and Rottschaefer (1983a). Recent discussions of the cognitive revolution in psychology generally appear in Gardner (1985) and Mahoney (1988).

5. Philosophical advances also contributed to the move from behaviorism to a cognitive approach in psychology, in particular, the critiques of analytic behaviorism, operationalism, and conventionalism (Erwin, 1978).

6. Of course, there is also Chomsky's (1959) well-known critique of the adequacy of behaviorist accounts of verbal behavior.

7. Critics have proposed four alternative explanations of the behavior in question: (1) the behavior is an adaptation response; (2) it is the result of sensitization; (3) the animals learn behaviors during the inescapable shock that are incompatible with escape, for instance, as we have seen, freezing; and (4) the inability to escape arises from a depletion of norepinephrine caused by the severity of the shock. Maier and Seligman reject these alternatives on both conceptual/theoretical and experimental grounds.

6

The social cognitive bases of morality

How do we learn to act morally?

6.1 THE COGNITIVE TURN IN PSYCHOLOGY

We have seen that an adequate scientific account of moral agency must include reference to our cognitive capacities. The questions of how we acquire and put into action our moral capacities cannot be answered satisfactorily by appealing only to evolutionarily based and learned, but noncognitive, moral capacities. In particular, we saw in Chapter 5 that the Skinnerian claim that the science of operant behavior is the science of values fails because it does not take account of the cognitive features of human agency. Our conclusion rested in large part on the findings of psychologists who have shown that cognitive factors play a role in the explanation of human behavior. This turn to the cognitive in psychology is not reserved for critics of behaviorist theory; it is part of what has come to be called the *cognitive revolution* in psychology, which has taken two forms. For want of better descriptions, I shall call one cognitive revolution the *representational revolution* and the other the *agential revolution*. The representational revolution focuses on human knowledge-gaining capacities and achievements. It has developed in the areas of perception, memory, imagery, language, thought, and problem solving. The agential revolution is concerned with issues of human action. It has emerged in the areas of learning, motivation, personality, social psychology, and abnormal psychology. Each in its own way stresses the cognitive capacities of humans both in coming to know the world and in learning to act in it. The representational revolution often draws on theories taken from computer science and information processing theories. In doing so, it has begun to move away from folk psychological conceptions of epistemic capacities and processes. In place of talk about beliefs and desires, it has begun to substitute notions like information, schemata, input and output buffers, and central processors. The agential revolution has to a larger degree retained folk psychological concepts of the cognitive. One branch of the agential revolution is a set of theories often called *cognitive social learning theories* or *cognitive behavioral theories* (Erwin 1978; Kazdin 1978; Dobson and Block 1988). Rooted to some degree in the behaviorist tradition, these theories have incorporated cognitive variables as independent variables into their theories of human agency. In this chapter, I explore the contribution of the agential revolution in psychology for understanding moral agency.[1] Specifically, I examine the contribution Albert Bandura's social cognitive theory of agency has made in answering questions about the acquisition and employment of moral agency.

The study of moral agency in scientific psychology, although rich with theories, seems thus far to have been unable to sort itself out (Kurtines and Grief 1974; Rest 1983; Hoffman 1988; Kurtines and Gewirtz 1988). No one theory has emerged that seems to have both strong empirical support and the ability to help us resolve some of the long-standing problems about moral agency. Indeed, the most prominent of these, cognitive developmental theories, although retaining many important features of our folk psychological conception of moral agency, have run into many empirical problems. But folk psychological and philosophical analyses, as well as scientific psychological investigations, have thus far left a theory of moral agency with many unresolved problems.

Although Bandura's social cognitive theory is concerned with human agency and motivation as such, I propose to examine its potential for explaining moral agency because I believe that moral agency, including moral motivation, stands a better chance of being accounted for within a more general theory of agency and motivation than one more narrowly focused.[2] Moreover, Bandura's theory of agency is an apt candidate because it is empirically well supported and theoretically well articulated. Thus one would expect that its prospects for extension would be brighter than those of other, less well established and elaborated theories. If one of the best of the cognitive social learning theories is found to be a poor candidate for solving some of the problems of moral agency, then the prospects for cognitive social learning theories generally and for the agential revolution in psychology are significantly dampened. In particular, I examine the fundamental challenge that cognitive social learning theories are unsuitable for explaining moral agency because their theories of agency are irrelevant to moral agency. That is, they are not about moral agency at all. If that is so, then the adequacy of their answers to the question of how we become agents and put our agential capacities to work will not apply to moral agency. This challenge arises in part from proponents of cognitive developmental theories of

126

moral agency that have dominated the study of moral agency since the 1960s.

In Sections 6.2 and 6.3, I focus on the problems of acquisition and action, assessing the relative successes of cognitive developmental theories and cognitive social learning theories in handling these problems. I lay out Bandura's social cognitive theory of agency and show how it can be extended to account for the acquisition and employment of moral agency. I conclude that although cognitive social learning theories are better able to handle the problems of acquisition and action, they face a major objection: that their accounts of human agency are irrelevant to moral agency. What emerges, then, for students of moral agency is a kind of *investigative dilemma:* The better theories of acquisition and agency, the cognitive social learning theories, seem to be irrelevant to the phenomena of moral agency, whereas the theories that seem to be relevant, the cognitive developmental theories, are unsatisfactory in their accounts of acquisition and action. In Sections 6.4 and 6.5, I explore this dilemma and resolve it in favor of cognitive social learning theories. I show that Bandura's social cognitive theory is relevant to moral agency, even by the standards proposed by its critics. Given the successes of cognitive social learning theories in general and Bandura's social cognitive theory in particular, it is reasonable to maintain that they are the best current candidates for accounting for moral agency.

6.2   COGNITIVE DEVELOPMENTAL ACCOUNTS OF MORAL AGENCY

Since the 1960s, the study of moral agency has been dominated by cognitive developmental theories. These theories appeared more plausible than either behaviorist or psychodynamic theories because the latter theories appeared to be incompatible with strongly held folk psychological views of moral agency. Appeals to only natural and social environmental factors in the explanation of behavior or to only noncognitive intrapsychic processes appeared to critics to neglect an essential aspect of morality: its cognitive motivation. Such limited causal accounts of human agency have lost much of the empirical support or explanatory power they seemed to possess in their heyday (Hoffman 1988). On the other hand, cognitive developmental theories, notably Lawrence Kohlberg's, not only seemed to fit with a folk psychological conception of moral agency and with some philosophical accounts, but also seemed to be empirically supported. Skinnerians and

Freudians could appeal to eliminative materialist principles and argue that the reliance on a folk psychological conception of moral agency is question begging. If either one of these views prevailed, the folk psychological image of our moral capacities and actions as cognitively based in moral principles would be radically false, and that image would be destined for at least the theoretical trash box. But both their scientific inadequacies and the strength of the folk psychological conception of moral agency have made the approach of cognitive developmental theories seem much more plausible than these classical competitors.

Kohlberg's (1976, 1981, 1984) is the most well known of the cognitive developmental theories. He has argued that moral development is the result of a progression through successively more adequate moral stages defined by moral cognitive structures that guide our actions. In its classical formulation it postulated six stages of moral development.[3] The stages are distinguished by the moral motivation that guides action. Stage I is egoistically motivated and based on external rewards and punishments. Stage II represents an enlightened egoism. Stage III makes social conformity within the circle of a person's immediate contacts the central motivational factor. In Stage IV that circle is expanded to the society in which a person lives. Stage V is characterized by philosophical utilitarianism in which the welfare of the greatest number of persons is primary. And the highest stage, Stage VI, considers not only the greatest happiness of all persons but what is due to an individual as a person. These stages are not age dependent, nor does Kohlberg claim that all persons reach the highest stage. In fact, most people, according to Kohlberg's research, arrive only at Stage IV. The stage features define aspects of the moral cognitive structures. The stages are *universal* insofar as they are found in all cultures. They are also *invariant* and *irreversible,* that is, there is no stage skipping and no movement backward in moral development. The stages are also *integrative* in that one stage builds on the previous one. A later stage logically contains the concerns of the previous one. Thus, for example, the conventional stages, III and IV, include the focus of preconvential stages, I and II, the self, within their larger scope of concern for the social group or society. Finally, the progression through the stages reflects an advance in *moral adequacy* from a least to a more adequate moral stance. All these features of moral growth show, according to Kohlberg, the guiding hand of nature in which the role of nurture, the influences of the social environment, play a facilitative but subordinate, nondeterminative role.[4]

Kohlberg's account has many attractive features. In particular, its emphasis on the cognitive component of moral development and the role of moral reasoning in moral development affirms a central feature of our folk psychological conception of moral agency, namely, both the necessary role of cognition in responsible behavior and the central role that intentions play in our estimate of moral behavior. But despite these advantages, it suffers from some significant problems. Here I can give only a sense of the current ongoing discussions about these problems; but in my judgment they are grave enough to make us search for an alternative to Kohlberg's theory.

Critics of Kohlberg have raised some empirical problems concerning the stage properties of universality, invariance, irreversibility, and integration (Mischel and Mischel 1976; Flanagan 1984). These properties, as I have indicated, reflect the marks of nature, of a built-in developing moral cognitive structure that flowers forth in individuals, provided that the facilitative input of the social environment is present. Unfortunately, the current evidence does not seem to be strong enough to support the existence of such stage properties. Indeed, Kohlberg (1984) has been forced to reconsider the existence of Stage VI and to withdraw the claim of cultural universality for this stage.[5] This retreat from the universality feature, if sustained, would be a major blow to his six-stage developmentalist program. Similar problems arise for the other features. There is evidence both for stage reversal and for exceptions to the invariant progress from stage to stage. Nor is it clear that integration exists except conceptually, and even then not for Stages V and VI. But perhaps most important, there is much evidence that there are no clear, distinct stages. Persons seem to be at several different stages and to mix stages in their reasoning.

These problems about the existence of stages of moral learning and the properties of such stages seem to indicate that Kohlberg's nature-emphasizing account of the acquisition of moral agency is problematic. They also suggest a reconsideration of the nurture-emphasizing approaches of the behaviorist tradition since the latter seems to have the tools for better accounting for the situationally varying and area-specific differences in moral belief and action that the data seem to reveal. Moreover, problems in the area of moral action, how we put our moral principles to work, incline us in a similar direction. The consensus of critics seems to be that even if one grants that Kohlberg gives us a plausible account of the types of moral judgment – a concession that many critics are not willing to make – he has not been successful in showing how moral principles lead to moral action (Blasi 1980; Flanagan 1984; Hoffman 1988).[6] On the whole, then, I conclude that cognitive developmental theories are relevant to moral agency, but are only marginally successful in answering the questions of moral action and seem to be inadequate in their account of the acquisition of moral capacities due to their reliance on a nature-emphasizing, stage conception of acquisition.[7] The problems with cognitive developmental theories, then, support a reconsideration of nurture-emphasizing approaches to moral agency like those of cognitive social learning theories.

6.3 COGNITIVE SOCIAL LEARNING THEORIES OF MORAL AGENCY

As we have seen, a major problem with nurture-emphasizing accounts has been their behaviorist colors. Learning in its behaviorist garb has been stripped of any cognitive strands, since for behaviorists, cognitions are, at best, the causally impotent effects of environmental influences. According to behaviorists, learning occurs when secondary conditioned stimuli are connected with primary built-in motivations by means of classical and operant conditioning. This view makes it possible to acquire a wide range of behaviors, depending on the prevailing social situations. But it also seems to imply that moral action is merely action in conformity with socially accepted standards and that moral learning involves only behavioral and affective accommodation to such standards. This picture of automatic agents gradually shaped into conformity with prevailing social standards also fares badly on empirical and philosophical grounds, besides being inadequate to our folk psychological experiences of moral agency. Thus, it seems, an adequate account of moral learning must on folk psychological, scientific, and philosophical grounds include cognitive learning. Consequently, social learning theories that give a causal role to cognition in effecting behavior appear to be a more promising, nurture-emphasizing alternative to Kohlberg than behaviorist theories of moral learning that omit the causal role of cognitive factors or treat them in an instrumentalist fashion, as neobehaviorists often do.

Cognitive social learning theorists show how such processes as anticipating, imagining, planning, monitoring, providing feedback, and cognitive restructuring, to name but a few, function in acquiring, maintaining, and changing behaviors.[8] However, they have not attempted to give a precise definition of cognition or describe the different kinds of cognition and cognitive processes that appear to be effective in bringing about action. They use a mixture of folk psychological and scientific theoretical terms drawn from information processing theories and elsewhere. As a result, there are a plethora of approaches, theories, and models that form only a loose unity around the theme of the importance of cognitions for the agential ends of acquiring, maintaining, enhancing, and changing behaviors (Dobson 1988; Mahoney 1988).

Social cognitive bases of morality

Albert Bandura's social cognitive theory and moral agency

One of the most prominent cognitive social learning theories is Albert Bandura's social cognitive theory. As a well-articulated and well-supported general theory of agency, it is a good candidate for accounting for moral agency. In the remainder of this section, I lay out Bandura's social cognitive theory of agency, showing how it accounts for the acquisition and employment of moral agency. Then, in the following sections, I examine and refute the charge that, despite their successes with respect to questions of acquisition and action, cognitive social learning theories generally and Bandura's social cognitive theory of agency in particular are irrelevant to moral agency.

In contrast to social learning theories of the behaviorist or neobehaviorist variety, Bandura's social cognitive theory, like other cognitive social learning theories, stresses the role of cognition in bringing about behavior (Bandura 1977b, 1986). In addition, his theory is characterized by the doctrine of *reciprocal determinism* and the theory of the *self-system.* By reciprocal determinism, Bandura means that environmental, behavioral, and cognitive factors interact with each other in effecting behavior (Bandura 1978b, 1986). The theory of the self-system postulates four major informational and motivational capacities and processes that play a central role in explaining human behavior. These concern (1) the formulation and imposition of individual goals and standards of performance; (2) self-monitoring; (3) self-reactive influences; and (4) self-efficacy judgments (Bandura, 1978b, 1982, 1986). The theory maintains that to explain successful and unsuccessful performance, including the regulation of thoughts and feelings, one must assume, in addition to the natural and social environmental sources of actions, not only behavioral and generically cognitive causes of behavior but also a set of cognitive processes that are self-referential.

Bandura's (1977b, pp. 17–22, 67–72; 1986, pp. 12–22, 116–22) social cognitive theory emphasizes the role of cognition in both learning and bringing about behaviors. Bandura uses both folk psychological and information processing conceptions in his discussion of these cognitive elements. Thus, Bandura holds, on the basis of an accumulation of research studies, that classical and operant conditioning are mediated by cognitive factors. Moreover, he contends that learning takes place not only by practice but also by modeling and symbolically. Thus, for instance, Bandura's (1977b, pp. 22–55; 1986, pp. 51–80) account of learning through modeling requires a rich assortment of cognitive factors: perceptual, attentional, and retentional. These include symbolic coding, cognitive organization, and symbolic rehearsal. For example, consider the cognitive mechanisms Bandura (1977b, pp. 155–8; 1986, pp. 377–89) invokes to explain how one can disengage her moral principles and their application from behavior either to excuse or to justify what would otherwise be an immoral action. According to Bandura, one can seek to justify the performance of an apparently immoral action by (1) using different or altered moral standards; (2) comparing the action favorably to other reprehensible alternatives; or (3) trying to change the understanding of the action by reclassifying it in a benign fashion, for instance, by euphemistic labeling. Immoral activity is also justified by blaming or attributing dehumanizing characteristics to the victim. Exoneration is sought by ignoring, minimizing, or misconstruing the effects of one's actions. Finally, Bandura contends that the person can find ways to diminish her responsibility for an action by claiming that she was only acting under orders or that the decision was not hers but the group's.

Besides the cognitive emphasis that Bandura's views share with other cognitive social learning theories, his views are characterized, as I have mentioned, by the theory of the self-system (Bandura 1977b, pp. 193–208; 1978b, pp. 344–58; 1986, pp. 22–46, chs. 8 and 9). Bandura has postulated that the self-system is a major cognitive and motivational factor in human behavior. In a number of different studies concerning a wide range of complex activities such as conquering phobias, inhibiting behaviors, regulating refractory behaviors, developing coping skills, reducing physiological stress, overcoming resignation and despondency, learning achievement strategies, increasing intrinsic interest, and choosing a career, Bandura and his colleagues have shown that the self-system plays an important role in the acquisition, maintenance, regulation, enhancement, and change of the behaviors, thoughts, and feelings necessary for accomplishing these activities and goals (Bandura 1977a,b, pp. 79–93, 128–58; 1982, pp. 122–47; 1986, chs. 8 and 9). Thus the self-system involves higher-level cognitive and motivational factors, factors involved in the regulation of lower-level cognitive and motivational capacities.

To assess social cognitive theory's relevance for moral agency, we need to consider the self-system in more detail. Two interrelated but distinct motivational subsystems make up the self-system: the *self-evaluative system,* constituted by competencies for forming self-standards and reacting evaluatively to efforts in attaining these standards, and the *self-efficacy system,* formed by competencies for self-efficacy judgments. The *self-monitoring* subsystem plays a necessary informational role in the operation of each of these subsystems. By means of the self-evaluative system, a person establishes goals and standards, both moral and nonmoral (Bandura 1977b, pp. 128–58; 1986, pp. 340–89). These goals and standards are acquired and maintained in a number of ways: by practice, modeling, and instruction. Their contents can reflect not merely general, cultural, or social norms, but also specific social influences and a person's own adaptation of general norms. Through evaluative self-reactions individuals acquire, maintain, regulate, and enhance their own behaviors by positively or negatively assessing their success in meeting self-prescribed standards (Bandura 1977b, p. 130; 1986, pp. 350–62). These reactions can take the form of either tangible or symbolic rewards or punishments or of self-approval or disapproval. The level of the self-motivation generated by self-reactions is a function of the type and value of the incentives and the nature of the performance standards. Although in Bandura's view self-administered tangible rewards or punishments do play a role in self-reaction, self-approval or disapproval can play a relatively independent motivating role. Thus, studies show that in areas that are particularly important for a person, a mere calculation of the cost-benefit ratios of the external consequences of behavior will not suffice to explain a behavior. Self-esteem effects must be included (Bandura 1976, p. 145; 1977b, pp. 143–5; 1986, pp. 230, 240–1, 254–5, 348–57, 366, 374–5).[9]

In order to understand the role of self-efficacy judgments, we need to distinguish between *perceived self-efficacy* and *outcome expectations.* The former refers to a person's judgments about her capacity to perform behaviors requisite to the achievement of certain outcomes. The latter has to do with expectations about whether a given behavior will achieve the outcome toward which it is aimed. Thus, outcome expectation is dependent on factors external to the agent. Moreover, self-efficacy judgments are not global claims about oneself and one's capabilities, but rather judgments about specific behaviors (Bandura 1977b, p. 84; 1986, pp. 422–49).[10] Bandura and his colleagues have found, in a number of studies with regard to a range of different behaviors, that perceived self-efficacy is a better predictor of actual performance than is previous performance. It forecasts the kinds of tasks that will be undertaken, in what situations, and the effort that will be expended in preparation for, performance of, and persistence in a task. Thus, perceived self-efficacy appears to be a major motivator of behavior.[11]

We can distinguish several stages of agency in Bandura's account. First, there is the *self as a cognitive agent.* At this stage, to put it in folk psychological terms, the self is conceived as an agent who operates on the basis of beliefs and desires whose actions are directed toward herself as well as others.[12] This first stage of the agency (which parallels the behavioral level of my model of moral agency) captures the first three functional criteria of moral agency, those of being a *cognitively motivated actor.* This stage reflects the emphasis on cognitive factors in social learning theories generally and distinguishes them from behaviorist positions in which the self is merely a locus of forces (Lacey 1979).

A second stage of motivation is that of the *self as a reflective cognitive agent* (corresponding to the reflective level of my model of moral agency). This second stage of agency includes a metalevel motivational system. One has higher-level beliefs and wants about one's lower-level beliefs and wants. The former influence behavior by influencing the latter. The higher-level wants are, in Alston's (1977) term, *self-evaluative standards.* These standards include but are not confined to moral standards. The self-evaluative subsystem of the self-system is a metalevel motivational source in Bandura's account. Its standards extend beyond the moral realm to other areas of competence, but they serve to guide both beliefs and desires and, thereby, behaviors.[13] I conclude that Bandura's social cognitive theory provides an account of moral agency that fulfills a fourth functional criterion of moral agency, that of a *morally motivated cognitive agency.*

Both the self-evaluative and self-efficacy subsystems suggest a third stage of agency, that of the *self as a self-referentially reflective cognitive agent* (corresponding to the fourth level of agency in my model of moral agency). There is a straightforward sense in which principles and standards of self-evaluation are self-referential insofar as these principles are meant to govern and motivate the actions, beliefs, and desires of the self whose they are. The principles and standards are also self-referential insofar as the self is active in their acquisition, maintenance, and application. But self-referentiality plays a further distinctive role in Bandura's theory of motivation and agency. For the goals and standards of behavior are not merely applied *by* the agent *to* herself; they also become *personal* goals and standards. They embody ideals, ways in which the person aspires to be morally, aesthetically, cognitively, socially, and so on. Self-satisfaction or dissatisfaction are motivating precisely because of their connection with these personal ideals. This aspect of self-referentiality differs from both the self-referentiality of applicability to the self and that of applicability by the self insofar as moral standards, for instance, can motivate in an impersonal way, moralistically or legalistically, or in a personal way, authentically. In the former cases, the moral or legal system can apply to a person and she can even apply it to herself. But in the latter case, the rules are the agent's in some fuller sense. This fuller sense seems to be what Hoffman (1988) has in mind by the highest level of moral internalization.[14] It also may require an evaluation of the standards themselves and thus implies, in accordance with Bandura's views, that there are various ways of acquiring, maintaining, and applying standards. However, the self-referential character of self-evaluative standards refers not to a global self, but to the ideal self in a specific behavioral area. The standards concern action, competencies, thoughts, desires, and feelings in a single area, for instance, playing a violin. As a result, the self of the ideal self is, in the first instance, an achieved, dynamic unity in a specific behavioral area. Most of Bandura's empirical work has focused on such specific areas. However, one can plausibly extend Bandura's concept of the self to fit both diachronically and synchronically more extensive personal unifications.[15]

The self-efficacy system confirms the presence of this third stage of agency but from a somewhat different perspective. We have been using simplified folk psychological conceptions of the factors involved in human agency referring only to beliefs and desires. A more adequate model must include capacities and competencies. If Bandura is correct, judgments about these factors are also motivating. Judgments of self-efficacy motivate not only behaviors, the objects of first-stage motivations, and desires and beliefs, the objects of second-stage motivations, but also the acquisition, maintenance, and application of self-standards, what we have designated as one object of third-stage motivations. For instance, negative self-efficacy judgments lead not only to behavioral failures but also to despondency and self-devaluation (Bandura 1977b, pp. 140–3; 1982, pp. 140–1; 1986, pp. 225–7, 359–60, 445–7). And self-efficacy judgments play a role in the selection of such self-standards as career choice and related competencies (Bandura 1982, pp. 135–6; 1986, pp. 430–5). Self-efficacy judgments, then, reveal both an element of reflexivity about capacities and competencies and an element of self-referentiality. The self involved here is the potential self, in contrast to the ideal self of the self-standards. Bandura's theory suggests, then, that human agency involves not only cognitive and reflective capacities but also self-referential capacities.

6.4   AN INVESTIGATIVE DILEMMA

Critics often contend that social learning theory accounts of moral agency are unsatisfactory because, despite their intentions, these theories are not about moral agency at all. Thus, even granting the apparent successes of cognitive social learning theories in accounting for how an agent acquires and puts into practice her agential capacities, cognitive developmental and social learning theorists of moral agency are faced with an *investigative dilemma:*

1. Even though cognitive developmental theories are relevant to moral agency, since they use folk psychological understandings of moral agency that seem ineliminable, especially the concepts of moral princi-

ples and motivation, they have failed to show that they can provide satisfactory solutions to the problems of acquisition and action.
2. On the other hand, the successes of social learning theories in accounting for these problems appear to be beside the point, since they are in important ways irrelevant to moral agency, no matter how appropriate they might be to other types of agency.
3. So, our most relevant scientific theories of moral agency, the cognitive developmental theories, are not highly confirmed empirically, and our highly confirmed scientific theories of agency, the social learning theories, are irrelevant to moral agency.
4. As a result, we are confronted with the uninviting options of pursuing relevant but largely unsubstantiated theories of moral agency or developing empirically successful accounts of agency that are irrelevant totally or in important respects to moral agency.
5. In either case, moral agency lacks understanding and explanation.

How have researchers reacted to this dilemma? Not surprisingly, some reject both of its alternatives. For instance, the social learning theorists Burton (1978), Nisan (1984), and Gibbs and Schnell (1985) argue for both the compatibility and the complementarity of the two approaches and thus for their integration. D. Kuhn (1978), a cognitive developmental theorist, contends that although both approaches as they stand are inadequate, further integrative work by both sides could result in a satisfactory unified theory. Cognitive developmental theorists who are critics of social learning theories attempt to dissolve the dilemma by maintaining some form of the irrelevance charge against social learning theories and arguing that the case against cognitive developmental theories either is not as devastating as made out or is remediable by modifications of cognitive developmental theories or their replacement by more adequate yet suitably cognitive and developmental theories.[16] For the most part, social learning theorists seem to have assumed the relevance of their investigations for issues of moral agency and thus have not addressed these charges directly. Taking up the cause of cognitive social learning theories generally and Bandura's social cognitive theory in particular, I challenge the charge of irrelevance and thereby seek to avoid the investigative dilemma by blunting its second horn.

There are several ways of meeting the charge of irrelevance. One could deny that the critics themselves have adequately pinned down the phenomenon of moral agency, that is, that they have failed to marshal the correct relevance criteria, and then go on to elaborate relevance criteria, showing how cognitive social learning theories meet them. Alternatively, one could accept the criteria suggested by the critics and argue that cognitive social learning theories fulfill them. I take the latter course. What justifies this approach? I suggest that one reason for accepting the critics' criteria is that at least some of them seem to represent entrenched folk psychological criteria. Of course, this will not satisfy skeptics of folk perspectives, whose concerns I shall address in the next chapter. However, if we focus on the current stage of the scientific psychological discussion of moral agency, folk psychological conceptions of moral agency still have an important place. Folk psychological relevance criteria, in particular, play a role with respect to the identification and description of moral agency. Specifically, a moral agent is understood to be an agent who is motivated by moral beliefs. This conception of moral agency is defensible on both folk psychological and philosophical grounds, and I argue that it finds support in cognitive social learning theories, in particular, Albert Bandura's social cognitive theory.

Questions of relevance concern issues of capturing the phenomenon in question with a proper description. As I have argued in our earlier discussion of relevance criteria, I do not take this task to be one of merely trying to explicate what is ordinarily understood to be moral agency, perhaps *the* folk psychological conception, or something that is accomplished by mere conceptual analysis. Although both folk psychological considerations and philosophical analyses do have something to say about what moral agency is, nevertheless, as is the case with all scientific investigations, attaining an adequate description is ultimately the result of finding empirical regularities and providing adequate causal accounts of these regularities. Indeed, the successes of cognitive social learning theory not only provide scientific support for the folk psychological conception of moral agency but refine it. Thus my contentions about relevance are, first, that cognitive social learning theories generally and Bandura's social cognitive theory in particular meet the functional criteria proposed by critics, specifically those embedded in the ordinary (and sometimes philosophically elaborated) notion of moral agency, that is, those of (1) agency, (2) cognitive agency, (3) cognitively motivated agency, and (4) morally cognitively motivated agency or, in other words, that a moral agent is a morally cognitively motivated agent. So, in this sense, cognitive social learning theories are as adequate with respect to relevance as the theories of their critics, some of which are cognitive developmental theories. Second, insofar as cognitive social learning theory descriptions are entrenched in more empirically supported regularities and explanations, they are more adequate to moral agency than their competitors. Finally, cognitive social learning theories bring to light a further, empirically supported criterion of relevance, one not currently advanced by competing theories, including cognitive developmental theories: that of moral integrity. According to this criterion, the agent conceives of herself as a moral person. Since this criterion is also entrenched in a general theory of agency that has substantial empirical support, I conclude that cognitive social learning theories are more relevant to moral agency than cognitive developmental theories.

6.5 RESOLVING THE INVESTIGATIVE DILEMMA

Proposed theories of moral agency can be either substantively or functionally irrelevant to moral agency or both. Some critics of social learning theories of moral agency, such as Lickona (1976) and Weinreich-Haste and Locke (1983), have found them to be substantively irrelevant, arguing that they are concerned with nonmoral matters such as conformity to social norms or the achievement of rewards and the avoidance of punishment. This objection fails because some social learning theories do in fact characterize moral actions in ways other than as actions in conformity with social norms or actions aimed at achieving rewards and avoiding punishment or as actions done for such motivations, for instance, those of Mischel and Mischel (1976, 1977), Kanfer and Karoly (1982), Karoly (1982), and Nisan (1984), and, as we have seen in some detail, Bandura's (1977b, 1986) social cognitive theory. Moreover, it seems to me that on both folk psychological and philosophical grounds, the substantive issue of what kinds of actions and intentions are relevantly moral is far from decided. First, some actions in conformity with social norms, for instance, those prohibiting homicide, are also in accord with moral norms. So, the substantive characterization of actions or intentions as in accord with social norms does not necessarily imply that these characterizations are nonmoral. In addition, although I do not support the view, one might even contend that moral norms are themselves based on social agreements. Finally, cognitive social learning theories are concerned with the explanation of such paradigmatically substantively relevant moral actions as prosocial behaviors. Thus, I contend that the charge of substantive irrelevance fails.

Working with the folk psychological account of moral agency, we have found that a moral agent is characterized functionally as a morally cognitively motivated agent. Thus moral agency involves four functional criteria: (1) agency, (2) cognitive agency, (3) cognitively motivated agency, and (4) morally cognitively motivated agency. I have argued that the agential cognitive revolution in psychology has demonstrated that cognition plays a role in human agency and thereby has shown the insufficiency of behaviorially based social learning theories of human agency.[17] Thus, I focus on the charge that cognitive social learning theories are nevertheless functionally irrelevant to moral agency because they cannot account for *morally cognitively motivated* agency.

One of the most persistent and philosophically acute critics of social learning theories of moral agency is the philosopher Thomas Wren (1982, 1991). He contends that social learning theories generally and Albert Bandura's social cognitive theory particularly are incapable of accounting for moral agency. Wren argues that social learning theories are unsatisfactory because they fail to account for the self-directed prescribing that he claims is essential to moral conduct. Although Wren does not explicitly distinguish issues of relevance and justification, I believe it is clear that he regards his rejection of social learning theories to be based on nonnormative criteria concerning the questions of what makes for morally cognitively motivated moral agency.

Wren identifies social learning theories as the most cognitivized portion of behaviorist learning theory. However, although social learning theories have adopted the current cognitive fashion, they remain, in Wren's view, fundamentally behavioristic, attempting to explain behavior primarily in terms of the "behavioristic credo, the Law of Effect" (Wren 1982, p. 410). In his 1991 critique of social learning theorists, Wren remains skeptical about their cognitive credentials, although he appears to grant that they posit more cognitive capacity than he had earlier thought. Nevertheless, he believes that the highly touted cognitive revolution in learning theories has yet to arrive and is currently no better than a "new twist" (1991, p. 25). Notwithstanding their newfound cognitivism, Wren continues to maintain that social learning theories fail to make the grade as theories of moral agency because of their failure to provide for genuine prescriptivity. Since they reduce moral demands to only one of many contending factors of equal standing in human motivation, they make moral motivation heteronomous.[18] Since I have assessed Wren's earlier critique elsewhere (Rottschaefer, 1991b), I focus here on his more recent rejection of cognitive social learning theories as capable of accounting for moral motivation.

Wren (1991) grants that there are a number of different socialization theories, some originating from social learning theories, which display varying depths of cognitivization. On a continuum, they range from (1) those that are noncognitive, employing no internal representations, through (2) those that are minimally cognitive, making use of some intrapsychic variables and involving some internal representations, to (3) those that are reflectively cognitive, involving self-regulation. The first group includes the radical behaviorist theories; the second group consists of social learning theories that make use of modeling and vicarious reinforcement; the third group contains social learning theories that invoke self-control by means of the delay of gratification or self-praise and self-blame. These latter are, in Wren's view, the most cognitivized of the social learning theories. Yet, despite this increasing degree of representational capacity and thus of *semantic* or *epistemological depth,* as he sometimes calls it, Wren maintains that social learning theories present morality and moral agency as heteronomous and are themselves externalist theories of moral motivation. The cognitions that they postulate do not provide moral motivation.

Wren offers two basic, related reasons for the failure of social learning theories to provide moral motivation that is cognitively based. First, they lack semantic depth; and second, their motivational resources remain contingently attached to the cognitions.[19] In Wren's view, moral cognition can motivate when it is capable of truth. Wren has a broad interpretation of truth. He does not restrict it to a correspondence sense, but includes senses compatible with truth as coherence and constructivist notions of truth (Wren 1991, p. 171). Indeed, he uses the term merely as a general one of cognitive assessment (Wren 1991, p. 170). Given this broad understanding of truth, that is, that the representational content is relatively adequate to what it represents, there seems to be no *in-principle* reason to contend, as Wren does, that social learning theories are incapable of the semantic depth required to meet his internalist account of cognitively morally motivated agency, provided that we grant, as Wren is willing to do, that some social learning theories are, indeed, cognitively based. For, as we have seen, it was precisely the positing of cognitive variables that enabled social learning theorists to better explain the verbal and nonverbal behaviors of their human subjects. A very plausible account of why such cognitive variables explain the phenomena under consideration better than even such internalized, but noncognitively interpreted, behaviorally based variables as learning history is precisely their representational adequacy. Indeed, arguably, realistically inclined interpreters of these studies can maintain that the adequacy conditions can and should be understood in terms of truth as correspondence. In any case, a proponent of the genuinely cognitive character of cognitive social learning theories can surely claim that the broad and minimal adequacy conditions for truth postulated by Wren are fulfilled. Thus Wren's first reason for denying that the cognitive variables postulated in social learning theories can provide moral motivation, namely, that they lack sufficient semantic depth, fails.

This brings us to the second major reason why Wren contends that social learning theories are in-principle functionally morally irrelevant. Despite their increasing cognitivization and their move away from behaviorism, these theories continue to postulate mechanical and externalist systems of motivation and thus promote heteronomous moral agency. Consequently, they cannot be about genuine morality and moral agency. Social learning theories continue to uphold mechanical systems of motivation because all motivation remains at the first level of agency, even though, in some instances at least, they postulate reflectively cognitive capacities and thus involve second-level beliefs (beliefs about beliefs). These theories conceive of motivation for moral actions to be either rewards and punishments dispensed by other social agents and institutions or self-administered external or internal rewards and punishments, the latter including self-praise and blame. Such motivations are heteronomous because the cognitions that interpret, express, and guide the action to be performed are only contingently connected with the motivations rather than being logically or necessarily connected with them. This contingency of connection marks the moral motivations postulated by social learning theories as externalist. Wren's internalist requirement, however, is a relatively relaxed one. He does not demand that motivating moral cognitions produce moral actions (Wren 1991, p. 14). Rather, the motivating moral cognitions must possess *some* causal efficacy (Wren 1990b, pp. 20–2; 1991, pp. 15–16).[20] According to Wren, externalist theories cannot provide such moral cognitive motivation because they postulate only contingent connections between the moral cognition and the motivation of the action. However, the connections are not contingent in the sense that they are arbitrary, that is, indeterminate or lawless, but in the sense that the rewards and punishments have no intrinsic or logical connections with the cognitions (Wren 1991, p. 38). Wren seems to be claiming that the moral motivation postulated by even the most cognitive social learning theories occurs through a more or less extended and complicated chain of reinforcers and punishers whose motivational power originated in pleasure and pain. Thus, even such self-regulated motivations as self-praise and blame gain their motivational power from hedonic sources, and even the most sophisticated cognitive representations motivates an agent only because of its associative links, through perhaps a long chain of reinforcers or punishers, with these primary reinforcers and punishers. The good as known does not motivate; and, perhaps, even what does motivate is not known. In contrast, Wren tells us concerning genuine moral motivation that "what I have in mind here is a general disposition or metamotivation, cutting across the historical and conceptual manifold of moral situations with their diverse sort of actions and moral principles, in such a way that the deliverances of moral judgment are understood by the agent as *exciting* as well as discriminating reasons for action" (1991, p. 9; emphasis added).

The plausibility of Wren's argument rests on the in-principle distinction that he postulates between the externalist/contingent link between moral cognition and moral motivation, on the one hand, and, on the other, the internalist/logical connection between the two that, on his view, an adequate theory of moral motivation requires. But there are several aspects of Wren's own account of cognitively based moral motivation that blur this distinction. We have seen that Wren interprets the internalist connection between moral cognition and motivation as a causal one. Given a person's belief about the moral rightness of an action, she is motivated – at least somewhat – to act on that belief. Wren allows us to understand the source of that motivation in either objectivist terms (the belief is about some real, objective moral values) or subjectivist terms (the belief embodies some moral project of the agent as a moral self). On the other hand, in externalist theories of moral motivation, moral cognitions lack intrinsic motivational power. They motivate only because of the associations they possess with genuine motivators, the various rewards and punishments paired with these cognitions over the course of the agent's learning history. Wren seems to be assuming that the cognitions ascribed to agents in social learning theories don't motivate either because they are not about the moral good or, even if they are, because they cannot motivate. Neither of these assumptions seems plausible. Once Wren grants the cognitivization of social learning theories, then there seems to be no in-principle reason why the cognitions postulated by social learning theories must function any differently than those postulated by internalists. They are capable in-principle of being about the moral good in either of Wren's senses of the term, objectively or subjectively, and thus capable of being morally motivating.

Indeed, supporters of cognitive social learning theories can appeal to two positive considerations for contending that the cognitive variables that they find necessary in order to explain the behavior of their subjects serve not only, to use Wren's terms, discriminatory but also exciting functions. First, recall that on both the radical behaviorist perspective of Skinner and the evolutionary perspective into which he incorporates it, pleasure and pain are not themselves primary reinforcers or punishers. Rather, they are cues and clues to the presence of reinforcers and punishers. The latter are aspects of things, persons, and circumstances that are, in the case of reinforcers, enhancing for the individual, a group, or the species. Thus Wren is mistaken in making pleasure and pain as such the in-principle source of motivation and value, even within a radical behaviorist position. The radical behaviorist tradition, as well as the evolutionary tradition into which it is incorporated, is objectivist in its account of both moral values and moral motivations. Second, we can plausibly understand the success of cognitive social learning theories in terms of the fact that the cognitive variables postulated by these theories to explain both the verbal reports and actions of their subjects are representational of both those things and their reinforcing and punishing properties; and, thereby, they are motivational.

Thus, Wren has offered no in-principle reasons for claiming that cognitive social learning theories generally are unable to meet the autonomy and intrinsic motivation conditions that he lays down as conditions for functional moral relevance. Indeed, there are positive reasons for claiming that such theories do meet these conditions and thus that they are in principle, at least, functionally morally relevant. In Wren's terms, cognitive social learning theories are in-principle internalist theories of morality and enable autonomous moral agency.

I conclude that Wren has failed to show that social learning theories cannot *in principle* provide for an adequate account of moral agency because they are not concerned with moral motivation. But we need not interpret Wren's claim so strongly. He may be claiming only that prominent social learning theories such as those of Bandura, the Mischels, and Kanfer and Karoly do *in fact* fail to meet the conditions of his model. And if some of the most prominent and successful of these theories prove to be morally irrelevant, we may have good empirical reasons, after all, to doubt the relevance of social learning theories for explaining moral agency. It will come as no surprise that I maintain that Bandura's social cognitive theory clearly meets Wren's conditions for autonomous moral agency, those of semantic depth and intrinsic moral motivation. Although I shall not argue the case here, since other prominent social learning theories, like those of the Mischels, and Kanfer and Karoly, propose similar models of agency, I contend that these theories, too, meet the requirements imposed by Wren. Hence, I maintain that Wren's thesis, interpreted as a specific critique of the ability of prominent social learning theories to deal with moral agency, also fails. In addition, as I suggested earlier, Bandura's theory has resources to meet the criterion of moral integrity speculatively suggested by Blasi, by Hoffman, and recently by Wren (1991). Indeed, and ironically, Bandura's social cognitive theory provides empirical support for it.

When we examine Wren's (1982, 1991) accounts of Bandura's views, we get a startlingly different picture from the one we laid out earlier. Although Wren (1991) has come to recognize the cognitive character of Bandura's social cognitive theory and classifies it, along with similar theories proposed by the Mischels, as well as Kanfer and Karoly, as an example of the most cognitivized form of social learning theory based theories of socialization, he nevertheless finds it to be proposing a heteronomous view of moral agency.

According to Wren (1982; 1991, pp. 57–66), one of the central features of Bandura's contribution to a social learning theory account of moral agency is the concept of imitative modeling and vicarious reinforcement. By means of imitative modeling one can learn behaviors vicariously by observation. Without performing in a similar fashion, one can learn to regulate one's own behavior by rewarding and punishing oneself in imitation of the rewarding and punishing behavior of others. But, in Wren's view, such learning achieves only heteronomous results. The cognitive component of such learning is merely an image of imitable behaviors and patterns of reward. It involves no evaluation of the basic motivating factors, the rewards and punishments, and thus remains a first-level cognitive factor; consequently, moral motivation does not have the requisite autonomy. Nor does it provide intrinsic motivation, since the motivational force of the self-reactive influence gains its strength, according to Wren, from its associationistic ties with primary affective reinforcers. Wren (1991, pp. 66–73) also recognizes that another contribution to the cognitivization of social learning theories stemming from theories like Bandura's is the incorporation of a model of self-control. Such models internalize the motivational system of the agent by putting some of the control of the reinforcers under the aegis of the agent rather than of external parties. Nevertheless, in Wren's view, this advance is not at all sufficient to allow Bandura's social cognitive theory to avoid externalism. For although the agent can to some extent control his or her behavior by delaying gratification, the fundamental motivation remains externalist, that is, concerned with pleasure and pain, although long-term gratifications are now taken into consideration. Finally, Bandura's social cognitive theory has taken one further step toward cognitivism by allowing a role for self-praise and self-blame and thus for internalized rewards (1991, pp. 82–6). But this last step is far from putting Bandura's social cognitive theory in the internalist camp, since, according to Wren, it represents, as developed by its proponents, a second-best reward system. According to Wren's interpretation of the social learning theories' proposals, a child comes by means of modeling to reward and punish his or her own behaviors by self-praise or self-blame, instead of being reinforced by parents. But in Wren's view, social learning theories fail to present an adult understanding of such self-praise and self-blame, and, indeed, given their theoretical position, are unable to do so. Thus, although the postulation of self-praise and self-blame reflects an advance in social learning theory accounts of cognitively based motivation, such motivations are just part of the more or less long chain of intermediate associated reinforcers that find their genuine motivational force in the externalist motives of pleasure and pain.

In my view, Wren has presented a very incomplete picture of Bandura's social cognitive theory and has misrepresented the small portion that he has presented. He does not mention Bandura's central theoretical position, the self-system, and refers to only one element of that position, self-rein-forcement, which, as I will show, he interprets in a way not representative of Bandura.[21] And although it is true that Bandura is a pioneer in studies of modeling or vicarious learning and its incorporation into social learning theory, Wren underestimates its cognitive character and fails to place it in the context of Bandura's overall views.

We have seen that Bandura contends, on the basis of a number of studies, that human agency is not only cognitive in character but reflectively and self-referentially so. Modeling, learning by practice and symbolically, all involve a number of higher-level cognitive capacities, such as symbolic coding, cognitive organization, and symbolic rehearsal. Thus, Wren has misrepresented Bandura's use of cognitive mediational factors by reducing them to an "image of the modeled behavior" and by failing to discuss, at all, Bandura's views on learning, including learning by practice and symbolic learning, both of which, according to Bandura, are cognitive in character. The cognitive mechanisms, mentioned earlier, that Bandura invokes to ex-plain how one can disengage her moral principles and their application from behavior, either to excuse or to justify what would otherwise be an immoral action, are especially significant instances of the possession of morally relevant cognitive capacities.

Wren's account also omits most of the central, distinctive parts of Band-ura's position. Besides the cognitive emphasis that Bandura's views share with other social learning theories, his views are, as we have seen, charac-terized by the doctrine of reciprocal determinism and the theory of the self-system. Thus, Wren inaccurately portrays Bandura's views by giving moti-vational ultimacy and primacy to extrinsic rewards and their associated imagistic representations. No such primacy exists in Bandura's view. In-deed, the theory of the self-system, based on a number of different, well-substantiated studies concerning a wide range of complex activities, demonstrates that the self-system is a major cognitive and motivational factor in human behavior and establishes the significant role that higher-level cognitions and motivations, including moral ones, play in effecting human behavior.

Given a full and adequate rendering of Bandura's theory of the self-system, it is easy to see that Wren's specific critiques fail. By means of the self-evaluative system, a person establishes goals and standards, including moral ones. These goals and standards reflect not merely general cultural or social norms but also specific social influences and a person's individual adaptation of general norms. The cognitive capacities constitutive of the self-evaluative and self-monitoring subsystems of the self-system, those involved in the reflective and self-referential levels of moral agency, are capable of achieving truth either in a correspondence, coherence, or constructivist sense. Thus they clearly meet Wren's minimal requirements on semantic depth. Furthermore, Bandura has established that, through their evaluative self-reactions, individuals acquire, maintain, regulate, and enhance behaviors. The level of the self-motivation generated by self-reactions is a function of the type and value of the incentives and the nature of the performance standards. Some of these incentives and performance standards are moral in character. Thus, the self-evaluative subsystem of the self-system provides higher-level motivational factors, distinct from lower-level cognitive and affective motivations, ones that have their own independent motivational input into the effecting of behaviors and that have some degree of control over these lower-level motivations. Wren's internalist criterion for autonomous moral agency is clearly met. That this is so is further demonstrated by the role of self-praise and blame in Bandura's account. We have seen that although Bandura allows that self-administered, tangible rewards and punishments motivate actions, he insists that self-approval or disapproval can play a relatively independent motivating role. Studies have established that a mere calculation of the cost-benefit ratios of the external consequences of behavior will not suffice to explain behaviors that are particularly important for a person. Self-esteem plays a necessary role in explaining such behaviors. The self-evaluative subsystem of the self-system, in terms of both its ability to provide reflective cognitive motivation and its self-referential cognitive motivation, is the source of such self-esteem effects. Thus it is clear that self- praise and self-blame are not, as Wren alleges, second-best reward systems. They provide the kind of intrinsic motivation demanded by Wren for autonomous moral agency.

Moreover, Bandura's theory also allows us to understand why the moral standards might themselves be motivating.[22] What motivates the fulfilling of the self-evaluative standards is the self-esteem or self-blame that is consequent on the fulfillment or nonfulfillment of standards rather than the fulfillment or nonfulfillment of some first-stage desire. Thus Bandura's account of moral agency includes the functional feature of moral integrity. Being moral has become a part of what it means for the person to be the kind of person she is or aspires to be. Thus, Bandura's work provides empirical support for a fifth functional criterion of moral agency, that of moral integrity.[23]

Thus I maintain that Wren's claim that Bandura's social cognitive theory does not have the resources to meet the conditions of his model of moral agency is incorrect. So, Wren has failed to establish that social learning theories generally and Bandura's social learning theory specifically possess "a truly important meta-ethical deficiency" (1982, p. 410). I have interpreted that deficiency to be one of intrinsic irrelevance due to a categorically unsatisfactory account of moral motivation. So, I conclude that Wren has failed to establish that Bandura's social cognitive theory is unacceptable because it is morally irrelevant. Moreover, his claim about social learning theories in general, that there are good reasons to believe that they are unsatisfactory because they are morally irrelevant, also fails. Positively, I have shown that one social learning theory, Bandura's, meets the criteria of moral relevance set down by Wren. Although I cannot argue for it here, I maintain that there are sufficiently relevant similarities between Bandura's views and some other prominent social learning theories, for instance, those referred to by Wren, the Mischels', and Kanfer and Karoly's, that my conclusion about meeting Wren's criteria of moral relevance also applies to them.[24] Moreover, the empirical support for Bandura's social cognitive theory, in particular for his theory of the self-system, indicates that there is some reason to assert the existence of a capacity for a fourth level of moral agency, the self as moral agent.

6.6 THE SELF-SYSTEM AND MORAL AGENCY

An examination of the state of the scientific question about moral agency and the prospects for cognitive social learning theories poses what I have called an investigative dilemma. Either students of moral agency must pursue a set of theories, cognitive developmental theories, which, although they appear from philosophical and folk psychological perspectives to be very relevant to moral agency, are empirically questionable on the scores of both acquisition and action; or students of moral agency have to pursue cognitive social learning theories, which, although they are relatively successful empirically in accounting for the acquisition of beliefs and motivations, and for actions on the bases of those beliefs and motivations, seem to be irrelevant to moral agency. Thus, in either case, it is unlikely that an adequate account of moral agency will be forthcoming. I have attempted to blunt the second horn of this dilemma. An examination of the general and specific critiques of the relevance of cognitive social learning theories to moral agency reveals a series of five functional criteria of moral agency: those of being an actor, a cognitive actor, a cognitively motivated actor, a morally cognitively motivated actor, and an integrally morally cognitively motivated actor. I have accepted the claim of critics that on the basis of either folk psychological or philosophical criteria these are not unreasonable functional criteria of moral agency, although they certainly are not definitively established and even though the last criterion ought to be understood only as a sufficient, not a necessary, condition for functional moral relevance. I have gone on to argue that cognitive social learning theories have not been shown to be generally unable to meet these criteria, and that in the specific case of one prominent social learning theory, Albert Bandura's social cognitive theory, the criteria are clearly met. Indeed, social cognitive theory provides a well-confirmed empirical theory to support these relevance criteria. Bandura's social cognitive theory and other similar cognitively based social learning theories answer questions of the acquisition of moral agency and how it is put into practice in a fashion superior to those of their main competitors, cognitive moral development theories. In addition, they spell out a set of criteria that enable us to answer the question of functional moral relevance in a specific fashion.

Thus, cognitively based social learning theories provide additional scientific support for the four-level model of moral agency that I have proposed. In Chapters 2 through 4, I have found some scientific support in Darwinian evolutionary theory, sociobiology, and biologically and psychologically based theories of development for a base level of moral agency constituted in part by evolutionarily based moral capacities. In Chapter 5, I provided evidence from behavioral psychology for another component of that base level: learned, noncognitive moral capacities. Thus, it seems reasonable to conclude tentatively at this juncture in our investigation that my four-level model of moral agency has received substantive scientific support and, consequently, that it provides a good model for an integrationist account of moral agency.[25]

The tentativeness of my conclusion at this point is due, first, to the fact that even if we have plausible scientific support for some answers to the questions of relevance, acquisition, and action, there are many facets of these questions that remain to be answered. Second, we still have to address the question of adequacy. But, most important, the scientific status of the cognitive revolution in psychology is open to question. I believe that a successful resolution of the investigative dilemma on behalf of cognitive social learning theories, and their continued successes in accounting for agency generally, lend credence to the genuineness of the agential revolution as a second cognitive revolution in psychology along with the representational revolution. But these successes do not of themselves resolve the issue of the scientific status of the cognitive revolution in psychology. Reductionist and eliminativist philosophers – enamored of the successes of the neurosciences and suspicious of the folk psychological framework of beliefs and desires in which cognitive social learning theories, including

Bandura's social cognitive theory, are often set – anticipate the reduction or elimination of such folk psychologically based theories, as well as those cognitive theories formulated explicitly and exclusively in information processing language. Thus, if my model of moral agency is inextricably linked with intentionalistic categories, like belief and desire, and if the scientific future is with the nonintentionalistic categories of the neurosciences, I face a reductionist predicament: To the extent that a folk psychological conception of agency is necessary for understanding moral agency, the understanding of moral agency will become scientifically inadequate; and to the extent that a model of agency is built on the nonintentionalistic theories of the neurosciences, it becomes irrelevant to moral agency. Thus, before I can declare victory for my four-level model of moral agency concerning the questions of relevance, acquisition, and action, based as it is on the intentionalistic categories of cognitive social learning theory, I must address the reductionist predicament. It is to that task that I now turn.

NOTES

1. Bower and Hilgard find not only that "social learning theory provides the best integrative summary of what modern learning theory has to contribute to the solution of practical problems" but also that it "provides a compatible framework within which to place information processing theories of language comprehension, memory, imagery and problem solving" (Bower and Hilgard 1981, p. 472). In their view, information processing theories provide the best current scientific characterization of the intentional categories of cognition and action. Thus, they conclude that "social learning theory may provide a basis of consensus for much of the research in the next decade" (1981, p. 472).
2. This view is not shared by all theorists. Compare the views of cognitive developmentalists like Rest (1983, 1984) and Turiel (1983).
3. Over the course of his very productive career, Kohlberg refined and revised his procedures and views extensively. The classical formulation of his theory of three levels and six stages of moral development can be found in Kohlberg (1981). A good account of the current revised theory appears in Kohlberg (1984).
4. Kohlberg has consistently contrasted his own interactionist views with those of both biological maturationism and environmentalism. However, in my view, his conception of interaction remains vague about the degree to which moral structures are the result of the influences of nurture and nature. My interpretation makes nature the primary factor. Owen Flanagan (1982) offers a different point of view.
5. Kohlberg now contends that his theory implies universality for all six stages but that there is as yet no empirical support for the presence of Stage VI moral reasoning in all cultures.

6. Kohlberg and Candee (1984) suggest as a connecting mechanism, judgments of responsibility; and they indicate that there are some correlations between moral stage, judgment of responsibility, and moral action.
7. Nor have Kohlberg's attempts to argue for the moral adequacy of his cognitive developmental theory been very successful. It is well known that both psychologists outside the cognitive developmental tradition and philosophers generally find these arguments fallacious (Alston 1971; Mischel and Mischel 1976; Flanagan 1984; Bandura 1986, 1988). I myself am not dissuaded of the prospects for a naturalistic justificatory approach implicit in some of Kohlberg's arguments and pursue this issue in Chapter 8.
8. Although the terminology is not fixed, I shall use the phrase *cognitive social learning theorists* to refer to those theorists, often coming out of the behaviorist tradition, who, while emphasizing social and natural environmental factors in accounting for human behavior, also postulate a role for cognitions in its explanation.
9. The very important role of self-censure or self-dissatisfaction is demonstrated very clearly in Bandura's (1977b, pp. 154–8; 1986, pp. 375–89) rich account of the strategies of disengagement from moral responsibility, mentioned earlier in the text, that people use to avoid self-censure and self-dissatisfaction.
10. These judgments vary on several dimensions important for performance, such as grade of difficulty of the task, generality of expectations, and strength of expectations. The judgments are complex, deriving by inference from cues about self-efficacy gained in actual performance, vicariously, persuasively, or emotively.
11. For critical discussion of Bandura's claims about self-efficacy, see Eastman and Marzillier (1984a,b) and Bandura (1984).
12. For the sake of simplicity and presentation, I have assimilated the conceptual components of Bandura's cognitivism to a folk psychological account in terms of beliefs and desires. In fact, his theory is a mixed one using both folk psychological and information processing conceptions of the mental.
13. Alston (1977, pp. 90–7) claims that he knows of no psychological theory that clearly distinguishes this second level of agency from the first level of agency. He believes that the empirical studies reporting the success of self-regulation in conflict situations fail to mark out such a level because they only report that the least desired option was chosen. He argues that such studies do not demonstrate any role for self-regulation independent of desires since the success of self-regulation may have been due merely to a shift in desire strengths among competing desires. Bandura's work, however, shows that both self-standards and self-efficacy judgments are operative in an agent's acting on the apparently less desired option, making it the more desired one. Alston's second problem with current psychological theories of motivation, including social learning theories, is the way they conceptualize motivational systems. They seem to reduce them to desires and exclude reason. Alston's complaint is another way, I believe, of pointing out that moral agency seems to require not just a cognitive

agent but a cognitively motivated agent. Although I believe that this criterion is fulfilled even by a first-level motivational system, the self-system clearly satisfies this criterion and thus answers Alston's second problem with psychological theories of motivation.

14. It is here that the conceptions of Bandura (1977b, 1978b, 1986, 1988), Blasi (1984), Damon (1984), Hoffman (1988), and Wren (1991) significantly converge. I have tried to capture this point in the functional criterion for moral agency of moral integrity. However, I am not persuaded that this criterion represents a necessary condition of moral agency. Moral integrity may be closer to the maximal end of the continuum of moral agency than the kind of moral agency we normally practice. Since I believe that the question of the kind of moral agency we have, like the question of what kind of rational agents we are, is a matter to be determined empirically, I do not want to decide that issue in an a priori fashion. More empirical work needs to be done on the general applicability of these criteria to moral action. In addition, it is not clear to me that the criteria of moral motivation and moral integrity identify exclusively moral criteria. In both Bandura's and Alston's accounts, self-standards are not necessarily moral. Perhaps, in the end, substantive criteria must be invoked to distinguish moral agency from other types of self-referentially reflexive agency. However, I do think it is significant that five students of moral agency, working from diverse theoretical approaches, have postulated a criterion of moral integrity. Moreover, it is especially ironic that critics of cognitive social learning theories who champion moral integrity as a requirement for being a moral agent can find empirical support for that requirement in the very sort of theory they find intrinsically irrelevant to moral agency.

15. It is interesting to consider to what extent personal unifications can be achieved within specific areas and between specific areas both across time and at a particular time in one's life. Such unifications, if possible, would also seem to be attainable in the moral domain.

16. Some of Kohlberg's disciples try to salvage his theory more or less intact. Others move further from Kohlberg. James Rest (1983, 1984) identifies cognitive developmental theories as theories of moral judgment and places them within a larger scheme of moral learning. William Damon (1984), Eliot Turiel (1983), and Augusto Blasi (1980, 1983, 1984) see the need, in different ways, to make more major modifications in Kohlberg's proposals. Turiel believes that a sharp distinction between the moral and conventional realms will allow for the emergence of a genuine cognitive developmental theory of morality. Damon argues that the moral realm must itself be broken up into specific domains to achieve an adequate cognitive developmental theory of morality. And both Damon and Blasi argue that a motivational self-system must be introduced in order to solve the problem of moral action.

17. It might be argued that Hans Eysenck's (1976) account of moral agency in terms of biological and conditioning factors fails to meet the functional criteria of cognitive agency. In addition, the prominent social learning theorists of moral

agency Justin Aronfreed (1968, 1976) and R. V. Burton (1984) both maintain the continuing importance of affective motivation, especially that derived from punishment, for effecting moral behavior. Although I shall not examine the issue here, it is arguable that such theories fail to meet the criterion of cognitively motivated agency.

18. One might object that Wren's conception of moral agency moves well beyond folk psychological conceptions. I shall not pause to consider this objection. Rather, I shall argue that cognitive social learning theories generally and Bandura's social cognitive theory in particular meet his formulation of the criterion of moral motivation, whether it is a folk psychological one or not.

19. According to Wren, both of these problems are rooted in the observer perspective intrinsic to Anglo-American scientific psychology. An internalist perspective, on the other hand, is free of these problems because it adopts a participant point of view. However, Wren does not push this separatist distinction to its full extent. Rather than urging a hard and fast distinction between explanatory and interpretive disciplines or between causes and reasons, he seeks a reflective equilibrium between metaethics and moral psychology. Nevertheless, it seems clear that it is metaethics, informed by internalist assumptions, that in the end dictates the terms of whatever agreement is to be negotiated between moral psychologists and moral philosophers (cf. Wren 1991, Ch. 1, and Wren 1990b).

20. As Wren exemplifies it, P1: "Eve believes that abortion is wrong" entails that P2: "Eve is at least somewhat motivated to oppose abortion." The term *entails* is to be understood both in a causal sense that ascribes some causal efficacy to the moral judgment embodied in P1 and in an expressive sense that moral judgment articulates a motivational structure possessed by the agent.

21. Although Wren refers to three of the four major books Bandura had authored at the time he was writing his 1982 critique of Bandura's views, he scarcely seems to use them. The best place to find a comprehensive account of Bandura's views before his most recent in-depth presentation (1986) was his *Social Learning Theory* (1977b), published well before Wren's 1981 account and cited by him. Other significant articles available to Wren were published in 1974, 1976, 1977, and 1978. By 1978, Bandura had set out the major features of his theory, reciprocal determinism and the self-system. Wren's 1991 book refers to Bandura's 1986 work, but it does not take into account Bandura's central theoretical construct, the self-system, and it continues to interpret Bandura's cognitivism and the role of cognitive motivators in an unwarranted minimalist fashion.

22. Wren (1982) seems to identify the prescriptivity of moral standards with a Kantian-like respect for the law. However, Wren (1991, pp. 18 and 161–5, for instance) relaxes his conditions for what constitutes a moral motivation, allowing both (1) formal reasons as, for instance, what he considers be an intrinsic motivation to be consistent and (2) the very general constraint that a moral motivation derives from what the agent considers to be most serious in his or her life. The latter seems to be more of a substantive constraint. Bandura's moral principles, on the other hand, could derive their prescriptivity from a variety of

formal or substantive sources – for instance, Aristotelian conceptions of virtue and ideal moral character. One need not adopt Wren's relatively more restrictive Kantian notion of what constitutes a motive reason as a moral reason. But whatever the content of the moral standards, Bandura's account seems to meet the functional condition invoked by Wren's model.

23. Following the lead of the cognitive developmental psychologist Antonio Blasi (1984) and the philosopher Charles Taylor (1977), Wren also makes the speculative proposal that the moral motivational system be grounded in a moral conception of the self, thus embracing what I have called the *criterion of moral integrity.* Ironically, it is Bandura who provides the empirical support for this proposal.

24. This is primarily so because Mischel and Mischel and Kanfer and Karoly invoke the notion of standards and self-referential cognitive capacities.

25. Although I cannot argue for it here, my model also instantiates a scientifically supported soft determinist account of freedom.

7

The neurophysiological bases of moral capacities

Does neurophysiology have room for moral agents?

Having solved the investigative dilemma facing scientific students of moral agency, thereby avoiding an ill-fated choice between a scientifically adequate but morally irrelevant account of agency and a morally relevant but scientifically inadequate account of agency, I concluded the previous chapter on a note of cautious optimism. The solution came by showing that a four-level model of moral agency based on Bandura's social cognitive theory of agency not only provides empirically well-grounded answers to major questions about how humans acquire and put into action their capacities as agents, but also more than meets the criteria for what counts as moral agency. However, my optimism is necessarily tempered by the realization that the investigative dilemma is only part of a larger challenge, what I call the *reductionist predicament,* facing any scientific naturalistic account of moral agency that takes as its goal to do justice both to the phenomenon of moral agency and to the scientifically established facts of agency. Reductionists and eliminativists are suspicious of the scientific adequacy of the psychological theories that I have invoked in the solution of the investigative dilemma. They question the scientific adequacy of theories that are formulated in terms of intentionalistic categories, whether of the folk psychological or of the information processing sort, or some combination thereof. One obvious source of their questions is the fact that, thus far, I have said nothing about the neurophysiological bases of the various capacities, both cognitive and noncognitive, that constitute my model of moral agency. What are the relationships between these capacities and the structures and functions of the brain on which, as a scientific naturalist, the integrationist must admit they are built? How do neurophysiological findings about the nature, structure, and function of the brain relate to the psychological and biological results used to construct my model of moral agency? In this chapter, I explore several major answers to these questions about whether, and if so how, our neurophysiology has room for moral

154

capacities. My goal is to determine to what extent my model of moral agency can withstand reductionist and eliminativist threats without lapsing from its scientific naturalistic commitments.

In Section 7.1, I lay out the classical account of reductionism to which identity theorists appeal in arguing for the reduction of psychological to neurophysiological theories and, thus, for the identity of the mental and the physical (or the mind and the brain). In addition, I discuss the revisions of that model due to attempts to make it fit historical cases of reduction. These revisions lead to a view of reduction as a continuum of relationships between basic and reduced theory. On one end of the continuum there are retentive reductions corresponding more or less to the identifications captured by the classical account of reduction; and, on the other end, eliminative reductions in which the "reduced theory" is discarded. Given this philosophical theory, I examine in Section 7.2 some of the reasons supporting a classical reduction and the mind–brain identity theory. I then discuss the recent functionalist insights that indicate that identity theory is probably an incorrect account of the relationships between psychological and neurophysiological theories and, thus, between the mental and the physical. The failure of identity theory can lead one in several directions. Eliminativists have taken its demise as an indication to push for an eliminative reduction and the elimination of psychological theories, whether formulated in folk psychological terms or other sorts of intentionalistic terms. Others, sometimes called *nonreductive physicalists,* have argued that the psychological supervenes on the neurophysiological and thus contend that the mental is neither reducible to the physical nor eliminable. I explore this latter option next in Section 7.3, arguing that although the notion of supervenience has the happy result of retaining the psychological level, something that appears necessary for any account of moral agency, it leads to a dilemma for the integrationist position that I have been urging. This *supervenience dilemma,* as I call it, seems to demand either psychological epiphenomenalism, that is, the causal inefficacy of cognitive factors, or a retreat to nonempirical explanations of agency. In Section 7.4, I resolve this dilemma by arguing for a materialistic emergent mentalism that rejects psychological epiphenomenalism and maintains that the cognitive and motivational capacities that constitute three levels of agency in my model are causally efficacious. In Section 7.5, I return to the eliminativist option and explore the reasons why eliminativists view a folk psychological theory of agency, or any scientific psychological theory of agency incorporating significant elements of folk psychology, as a prime candidate for elimination and replacement by a more adequate scientific theory of our capacities as agents, more than likely a neuroscientific one. I conclude in Section 7.6 by reviewing the status of the reductionist predicament, focusing on the choice between eliminativist and emergentist positions with respect to our capacities as moral agents.

7.1   THE REDUCTIONIST PROGRAM: IDENTITY THEORY AND ELIMINATIVE MATERIALISM

What are the relationships between the cognitive and motivational psychological theories that form the basis of the behavioral, reflective, and self-referential levels of my account of moral agency and the neurophysiological theories that account for the structure and functioning of the brain? Surely any self-respecting scientific naturalistic account of moral agency has to say something about these relationships. Indeed, in raising the reductionist predicament, reductionists and eliminativists challenge the scientific adequacy of my model of moral agency, arguing either that the cognitive psychological theories on which my account of moral agency in large part rests will, in the former's view, be reduced to more basic neuroscientific theories or, in the latter's view, be eliminated. In order to meet this challenge, I first have to lay out in more detail the positions of the reductionist and the eliminativist.

Without really saying so, we have gotten into the modern version of the old-fashioned philosophical problem of the mind and the body. As that problem comes to us, at least from the time of Descartes in the sixteenth century, the question has been framed in terms of what the relationship is between the mind and the body. The standard classical answers to that question have been monistic and dualistic. The two most prominent forms of monism are materialism and idealism. Materialism is the metaphysical view that everything is matter; idealism is the metaphysical view that everything is immaterial. Dualism is the metaphysical view that mind and body (or brain) are two distinct substances. It is surely an understatement to say that idealist and dualist solutions to the mind–body problem have fallen on philosophical hard times. On the other hand, idealism and dualism seem to have at least held their own in ordinary and religious thought. Nevertheless, I focus my attention more narrowly within the materialist tradition, for it is there that we meet reductionists' and eliminativists' solutions to the mind–brain problem. In the context of the mind–body problem, reductionists have come to be referred to as *identity theorists* since they claim that the mind and the brain are identical. They maintain that this ontological claim can be demonstrated by the reduction of theories about our mental capacities and their functioning to theories about the brain and its functioning. To complete our story and see how a discussion of the mind–brain problem affects our pursuit of questions about moral agency, I bring in two other materialist answers to the mind–brain problem besides those of the identity theorists and the eliminativists: functionalist and emergentist solutions. Let's start by considering the identity theorists' and eliminativists' respective points of view.

To understand the difference between identity theory and eliminative materialism, one has to understand what *reduction* means. The identity theorist believes that mental phenomena can be identified with physical phenomena because theories about the mental can be reduced to theories about the physical. More specifically, folk psychological theories about the mind or theories in cognitive science about mental capacities can be reduced to neurophysiological theories about the brain. In order to understand the classical account of theory reduction, one first needs to know something about the *deductive–nomological model of explanation* developed by logical empiricist philosophers beginning in the 1940s and 1950s (Hempel and Oppenheim 1948; Nagel 1961; Hempel 1965; and Kitcher and Salmon 1989). According to the deductive–nomological model of explanation, an explanation has the form of a valid deductive argument.[1] A valid deductive argument is one such that if its premises are true, then its conclusion must be true. The explanans, the premises of the argument, contains the explanatory information, and the conclusion (the explanandum) contains what is to be explained. So, for example, we can explain an event or a state of affairs by means of a general law and initial conditions. If we want to know why the bar has expanded, then an explanation is given by a valid deductive argument of the following sort:

All copper expands when heated. (General Law)
This bar is copper. (Initial Condition)
This bar was heated. (Initial Condition)
So, this bar expanded. (Explanandum)

Similarly, we can explain one law by means of other laws. For example, disregarding the complexities:

If Newton's law of gravity and three laws of motion are correct, then
    Kepler's three laws about planetary motion are correct.
But Newton's laws are correct.
So Kepler's laws are correct.

Given this understanding of explanation, we can now consider the classical model of reduction, the *deductive–nomological model of reduction*

(Nagel 1961; P. S. Churchland 1986). Reduction is, first of all, a relationship between two theories: One theory, the reduced theory, T_R, is reduced to another, more basic theory, T_B, if and only if:

(a) *Derivation Thesis:* T_R is deduced from T_B plus some extra stuff (A), for instance, statements of initial conditions, and

(b) *Identification Thesis:* The relevant terms of T_R are identified with those of T_B.

Many have cited the reduction of phenomenological thermodynamics to statistical mechanics in the kinetic theory of gases as a reduction of this sort. Thus considering the conditions in reverse order:

II. *Identification Thesis:* The terms of the theory to be reduced, T_R, those that refer to the phenomenological, that is, perceptible, macroscopic properties of gases, their temperature and pressure, are identified by means of the correspondence rules or bridge principles with those of the reducing theory, T_B, whose terms refer to the imperceptible microscopic properties of the particles, their average energy and momentum, respectively. *The relevant terms of T_R are identified with those of T_B.*

I. Given the fulfillment of the Identification Thesis, then, one can fulfill the *Derivation Thesis.* One can deduce from the kinetic theory of gases, T_R, using some other stuff, including here the correspondence rules, some laws of phenomenological thermodynamics (for example, $PV = nRT$. *T_R is deduced from T_B plus some extra stuff (A)).*

Let us examine a reconstructed example of a proposed reduction of a psychological theory to a neurophysiological one. Consider Freud's project for a scientific psychology as applied to a case of hysteria and its explanation (Fancher 1973, Chs. 2 and 3). One symptom of hysteria is an inability to use or move a limb. For instance, suppose that I am unable to move my arm. There is no immediately detectable physical source of my symptom. I also have other symptoms, such as periodic outbursts of uncontrollable laughter, that, together with my inability to move my arm, constitute the hysterical syndrome (HS). First, we explain HS using the deductive–nomological model and relating the observable behaviors of HS to unobservable psychological factors, repressed ideas (RI). So, we have if RI, then HS. We have a *psychological level of explanation.* Next, we postulate bridge laws/rules of correspondence to relate the psychological and neurophysiological levels. Freud, whose investigations and speculations occurred well before many of the recent advances in the neurosciences, talked about a Q distribution in neuronal states. We might think of this nowadays as a configuration of neuronal firings and neurotransmitter levels. Freud postulated that RIs are identical to certain Q distributions in the neurons: RI = Q. So, we have a *bridge law relating the psychological and the neurophysiological.* Freud also anticipated a further set of bridge laws linking the neurophysiological with the chemical and the physical. So, he postulated another set of bridge principles/correspondence rules linking those levels. He characterized the physical/chemical levels together in terms of energy, E. So, we have Q = E. Thus, we have a *bridge law relating the neurophysiological with the physical/chemical.* Consequently, we can explain HS in terms of E by means of the bridge principles and the psychological law, that is, if E, then HS. This example illustrates the reduction of a psychological theory to a neurophysiological theory and the latter to a chemical/physical theory in the classical deductive–nomological sense.[2]

The logical empiricist deductive–nomological model of reduction has been criticized from various perspectives. Some who think that there is still a lot to it nevertheless believe that it requires some revision. I shall lay out a detailed account of this revised model (P. M. Churchland 1984, pp. 26–7; P. S. Churchland 1986, Ch 7; Schaffner 1993). A problem with the deductive–nomological model of reduction is that it does not fit many actual cases exactly. A major problem is that the old theory, the T_R, cannot often, if ever, be deduced from the new theory, T_B, without modifying T_R. At best, only a corrected version of the old theory, call it T_{R*}, can be deduced from T_B and the extra stuff, A, for example, the bridge principles, auxiliary theories, and initial conditions. So, it seems necessary to revise the deductive–nomological model of reduction as follows:

Logical Empiricist Version of the Deductive–Nomological Model of Reduction: T_B + A leads deductively to T_R.
Revised Deductive–Nomological Model of Reduction: T_B + A leads deductively to T_{R*}, and T_{R*} is more or less analogous to T_R.

There are some cases where T_{R*} mirrors T_R quite closely or is very analogous to it, for instance, the reduction of light to electromagnetic radiation, sounds to mechanical waves of the medium, and the temperature of a gas to the kinetic energy of particles. But there are other cases in which T_{R*} mirrors T_R less closely and indeed corrects T_R. The correction of Newton's laws by Einstein's theory of special relativity is one example often referred to.[3] We need not go into an analysis of this case; but, roughly, here's how it is understood by advocates of the revised model. If Einstein's theory of special relativity is correct, then Newton's assumption that mass is a constant independent of velocity is not correct. The amount of mass of a body depends on its velocity. So, what is deduced from Einstein's theory of special relativity is not Newton's laws, since they assume mass to be a constant independent of velocity, but something more or less analogous to Newton's laws. But the history of science tells us even more. It tells us that sometimes, T_{R*} is very different from T_R. For instance, the old theory of heat, the caloric theory, is very different from the new corpuscular/kinetic theory; the old theory of combustion, the phlogiston theory, is very different from the new one that involves oxygen; the old Earth-centered, crystalline spheres of Aristotelian astronomy are very different from the Copernican sun-centered system; and the old theory of psychosis, the witchcraft theory, is very different from the new ones based on neurophysiology.

What the history of science reveals, then, is a range of possible relationships between T_{R*} and T_R and thus a range of relationships between T_B and T_R. We can think of it this way:

I. T_B + A lead deductively to T_{R*} and
II. either
    a. T_{R*} is very similar to T_R, so we have a *retentive reduction,* that is, we identify some of the terms of T_B with those of T_R, and so we retain the latter (e.g., the identity of mental and physical terms as proposed by identity theorists), or
    b. T_{R*} is somewhat similar to T_R, so we have a *partially retentive reduction,* that is, we find that the terms of T_B have some connections to those of T_R and are, in part, retained, or
    c. T_{R*} is more dissimilar to T_R than similar, so we have a *partially eliminative reduction,* that is, many of the terms of T_B are dissimilar to those of T_R and we eliminate many of the latter, or
    d. T_{R*} is completely or almost completely dissimilar to T_R, so we have an *elimination* of the old theory, that is, many of the terms of T_B are completely different from those of T_R and we eliminate them.

Notice that in all cases T_B + A should explain why T_R explained what it did to the extent that it did and why it did not to the extent that it did not. Of course, T_B must be a justified theory, and the stuff in A must be justified, and both must be in the domain of the subject matter of T_R.

At present, of course, we don't know what will happen with respect to the mind–brain issue. Bets are still open. But consider a future scenario on the relationships between the neurosciences and psychology: T_B (some future neuroscientific theory) + A (bridge principles, auxiliary hypotheses, etc.) lead deductively to T_{R*} and T_{R*} is most similar to T_{R5}, a scientific psychological theory (SPT) formulated in nonintentional terms. Moreover,

T_{R*} is less similar to T_{R4}, an SPT formulated in intentional terms, but not of the sort employed in folk psychology, and still less similar to T_{R3}, an SPT formulated in intentional terms and retaining some folk psychological terms. Finally, it is still less similar to T_{R2}, an SPT formulated in intentional terms and retaining important elements of folk psychology, and is least similar to T_{R1}, folk psychology. Eliminativists are wagering, on the basis of the history of science and what we currently know, that (1) T_{R5} will probably be the form of the SPT of the future and T_B and T_{R5} will coevolve as top-down and bottom-up strategies are simultaneously employed; (2) T_{R1}, folk psychology, will be eliminated; (3) T_{R4}, T_{R3}, and T_{R2} will be less successful psychological theories than T_{R5} and thus will gradually be rejected in favor of the latter. On the other hand, the identity theorists are betting on T_{R1} (folk psychology) or on one or the other of the SPTs being reduced to T_B. That is, T_{R*} will be very similar to or identical with folk psychology or one of the SPTs. The functionalists and emergentists, whose positions I examine shortly, deny that a reduction is going to occur; that is, there will be no T_B to which some future comprehensive scientific psychological theory will be reduced. The latter will remain, in some sense of the term, an autonomous theory.

7.2 THE NEUROPHYSIOLOGICAL BASES OF MORAL AGENCY: REDUCTIONISM AND THE FUNCTIONALIST RESPONSE TO REDUCTIONISM

The reductionist predicament poses an unwanted trade-off for the integrationist who is trying to develop a scientifically based account of moral agency: be scientifically adequate and give up talking about moral agency or talk about moral agency but give up being scientifically adequate. Why does the reductionist think that these two options are the only ones that the integrationist has? Why cannot the integrationist have her cake and eat it? Why is it fruitless to try to develop a scientifically adequate account of moral agency?

To answer these questions, we need to find out what is wrong with the cognitivism that the integrationist contends is necessary for an adequate account of the acquisition and activation of a person's agency, both moral and nonmoral. For the most part, thus far, I have discussed moral agency in terms of beliefs and desires. Bandura's conception of agency was laid out folk psychologically in terms of a set of levels of beliefs and desires. We also know that he uses the information processing terminology of contemporary cognitive science. What's so scientifically inappropriate about either version of cognitivism that it must be reduced to neurophysiology or eliminated? In this section, I examine the retentive reductionist side of this challenge and the functionalist response to it, leaving the discussion of the eliminativist challenge to a later section.

First, since I am a card-carrying materialist, it should be clear that the integrationist advocate of my model of moral agency is in complete agreement with reductionists that our cognitive capacities are based in our neurophysiological ones. Both parties agree that the brain is necessary for the functioning of the mind. A cursory examination of the evidence makes this clear (Bloom and Lazerson 1988; Changeux 1985; P. S. Churchland 1986). Consider first the sensory and behavioral capacities necessary for moral agency. Neurophysiologists and psychologists have identified what is called the *sensory-motor cortex* as the basis for these capacities. Similarly, the neurophysiological structures of our perceptual systems, the pathways from stimulus input to higher levels of integration, have been traced, especially for the visual system. Centers for the apprehension, comprehension, and use of language have been identified. Very detailed accounts of habituation and sensitization in simple organisms have been given in terms of neural pathways, the transmission of electrochemical signals and neurotransmitters, and there are promising proposals for similar explanations of respondent and operant conditioning. There is much evidence to link the limbic system with our emotional capacities and responses. In addition, very recently some promising evidence has been provided for locating our capacities for personal, social, and moral judgment. Hanna and Antonio Damasio and their colleagues (Damasio 1994; Damasio et al. 1994) have argued that the classical case of Phineas Gage, as well as patients with lesions similar to that of Gage, can be understood in terms of impairments to both the left and right prefrontal cortices that cause a defect in rational decision making with regard to personal, social, and moral matters and with the processing of emotions necessary for these decisions.

All of this evidence, and much more, provides overwhelming support for a materialist account of our cognitive capacities. Our cognitive capacities require the neurophysiological structures and functions of the brain for their existence, maintenance, and functioning. However, the retentive reductionist contends that this evidence points to and supports identity theory; psychological variables are to be identified with neurophysiologically identified variables, and psychological laws are to be derived from neurophysiological laws. In a global fashion, we might conceive of the input, mediational, and output capacities constitutive of the perceptual, cognitive, motivational, and behavioral capacities of moral agency as identifiable with a complex of (1) neuronal networks and dynamic configurations of (2) neuronal firings and (3) neurotransmitters (Changeux 1985). Although most retentive reductionists admit that such a reduction has not yet been achieved, they argue that all the evidence points to it and that it can be expected as part of future scientific achievements, just as there have been successful reductions, so they claim, of biological to chemical theories (e.g., the reduction of classical Mendelian theories of genetics to molecular biological theories of RNA and DNA) and chemical theories to those of physics (e.g., theories of the chemical bonding properties of atoms to quantum mechanical and quantum dynamic theories).

Nevertheless, the emergence of cognitive psychology and artificial intelligence research, along with the development of functionalism in philosophy of mind, has led to a widespread rejection of the adequacy of the reductionist ideal for understanding the relationships between the mental and the neurophysiological. Briefly, the functionalist view goes something like this. The mental should be conceived of in terms of a set of roles. The meaning of a concept (or term) or proposition (or sentence) is determined by a set of other meanings that are linked by a set of inferential relationships. Thus, the mental involves a set of causal inputs, a resulting set of internal causal networks, and a set of causal outputs. Wilfrid Sellars (1968, Chs. III–V; 1979; 1980; Rottschaefer 1987) developed an analogy between mental processes and chess. We can conceive of chess as a set of rule-governed activities. These rules govern the movement of certain pieces on a board that is structured in a certain fashion and with certain end results to be achieved. Sellars opines that one could play ordinary chess on an ordinary chessboard or one could play what he lovingly calls *Texas chess* using the counties of Texas as one's board and Volkswagens for pawns, Cadillacs or Lincolns for kings and queens, and so on. Bruce Aune (1967) added that it is, of course, possible to play flea-bitten chess on the hide of some poor rat! The application to mind is straightforward. If the mental is constituted by a set of causal relationships of input, internal interactions, and output, then those relationships could be instantiated in a large, perhaps indefinite, variety of materials. After all, one can build a bird feeder out of many different things and it will still be a bird feeder. So, too, mental phenomena need not be made of just the neuronal wet-ware of humans. They could also be made out of the silicon chip hardware of computers, the Mars-based materials of Martians, the spiritual material of the archangel Gabriel (!), and so on.

If the functionalists are right, then the classical reductionists are wrong. The mental is not reducible to a particular type of physical material. The hopes of the retentive reductionists, at least with respect to the reduction of intentional mental kinds to neurophysiological kinds, seem to be dashed. At best, each instance of mentality is identified with some instance of the physical. There is only token-token identity, not type-type identity, as the classical reductionists had maintained. This understanding of the relationship between the mental and physical has been explicated in terms of the notion of supervenience. I now examine the notion of supervenience to see how it is used to explicate the irreducibility of the mental. That will enable us to see how a persistent reductionist can herself make use of the supervenience relationship in an attempt to turn the tables on the antireductionists.

7.3 THE BLESSINGS AND CURSE OF SUPERVENIENCE

Supervenience has come into its own as a distinctive and important philosophical concept (Kim 1990b). Although it was introduced into contemporary philosophical discourse in connection with morality, discussions of its applications have extended to philosophical psychology and to ontological and epistemological issues centering on reductionism and determination. Supervenience has two major features: multiple instantiation and supervenient determination, roughly, the realization of a higher level of reality in a lower level and the effecting of higher-level phenomena by lower-level phenomena.[4] These two features seem to supply materialists with a way to maintain their materialist commitments while abandoning restrictive and perhaps empirically unrealistic reductionistic views.[5] Supervenience may also strike a responsive cord with integrationists who make use of evolutionary theory in their account of moral agency, since it appears to capture by means of the notion of multiple instantiation the evolutionary idea of emergent levels of reality while at the same time implying the continuity of nature via the concepts of instantiation and supervenient determination.

The functionalist understanding of the relationship between the mental and physical has been explicated in terms of the notions of determination and multiple instantiation as follows: (1) one level of reality is instantiated or realized in a lower level of reality and (2) that lower level determines the upper level in the following manner: (a) If two entities on level 1 (L1) have exactly the same subvenient base phenomena on level 2 (L2), they will have exactly the same supervenient phenomena. For example, if Paul and Pat have the same set of neurophysiological patterns in their associative cortices, then they have exactly the same thought, for example, that eliminative materialism is true. However, (b) two entities on L1 with exactly the same supervenient phenomena need not have exactly the same subvenient base phenomena on L2. For example, if Paul and Pat are each thinking the same thought, that eliminative materialism is true, then they need not have the same neurophysiological patterns in their associative cortices. Subvenient bases determine upper-level supervenient phenomena, but different subvenient bases can bring about the same upper-level supervenient phenomena.[6]

Returning to moral agency, what implications does supervenience have for an integrationist who is committed to a complex account of moral agency that includes four functional levels? On first inspection, it would seem that the prospects are good. Supervenience appears to account for both the continuity of moral agency with nonmoral phenomena and its emergence, and thus distinction, from nonmoral phenomena. Moral phenomena supervene on nonmoral phenomena in the sense discussed earlier: Moral phenomena are instantiated in nonmoral phenomena in such a way that given that the subvenient bases of two moral phenomena are identical, then these moral phenomena are identical; but two moral phenomena of the same sort need not have the same subvenient bases. Thus, with respect to ontological questions, the integrationist can maintain that there is a distinction between emergent moral phenomena, including moral agency, and their subvenient nonmoral bases, and at the same time can hold on to a naturalistic continuity between moral and nonmoral phenomena. The former goal is achieved because, unlike a reductionist, the integrationist does not identify moral and nonmoral phenomena. This is because of the phenomenon of multiple instantiation.[7] The goal of continuity is achieved because of the occurrence of determination: Subvenient bases determine supervenient properties. Explanatory adequacy seems assured because of the supervenience of the higher levels of agency on lower levels. As in the case of ontology, so too with respect to explanation: Multiple instantiation provides for emergence by means of the notion of levels, and supervenient determination ensures continuity. Thus, supervenience seems to be a way for an integrationist to have her cake and eat it.

But a second look gives one pause. Two questions arise. Doesn't supervenience also entail epiphenomenalism? And isn't epiphenomenalism incompatible with moral agency, if not also with emergence? Epiphenomenal properties and entities are causally inert. Thus, even though the epiphenomenalist admits distinct levels of supervenient entities and properties, those entities and properties exercise no causality. The supervenience relation leads to the following dilemma with respect to explanatory emergence and, consequently, to the functional moral relevance of the integrationist account of moral agency: (1) Either moral actions supervene on nonmoral events or they do not. (2) If they do, then moral actions are determined by nonmoral events. But such determination entails explanatory ethical epiphenomenalism since it renders the higher, moral level explanatorily inert. (3) On the other hand, if moral actions do not supervene on nonmoral events, then they are scientifically inexplicable and unjustifiable. If so, they must belong to some nonnaturalistic or supernatural realm. (4) In either case, an integrationist account of moral agency is futile. Thus it seems that supervenience is a two-edged sword. Although it may save an integrationist account of moral agency from reductionism, it seems to force it into explanatory epiphenomenalism. So, supervenience seems to secure explanatory, including evolutionary, continuity, but at the expense of depriving moral agents of any genuine agency.

7.4 RESOLVING THE SUPERVENIENCE DILEMMA

Since I want to retain the benefits of supervenience, I will concede the second horn of the dilemma and blunt its first horn. Call an explanation of moral phenomena by nonmoral phenomena a form of supervenience explanation. In general, supervenience explanations are explanations of supervenient facts, events, kinds, and the like by subvenient facts, events, and so on. The *exclusive* supervenience explanation thesis states that supervenience explanations are necessary and sufficient for the explanation of higher-level phenomena. As applied to moral phenomena, it implies that lower-level theories about nonmoral phenomena, for instance, in terms of neurophysiological events, are all that are needed to explain moral phenomena, that is, moral actions, virtues, moral goods, and so on. Thus, I reformulate the first horn of the dilemma as follows: (2′) The doctrine of supervenience leads to the exclusive supervenience explanation thesis, which entails ethical epiphenomenalism. I shall try to show that the doctrine of supervenience does not imply the exclusive supervenience explanation thesis. To do this, I will consider an important representative argument for the exclusive supervenience explanation thesis fashioned by Jaegwon Kim.

Kim (1984b, 1989, 1990a, 1992b, 1993) has developed the exclusive supervenience explanation thesis and its epiphenomenal implications for mental causation most thoroughly. I examine a reconstruction of his argument to show that it is problematic with respect to both mental causation generally and moral agency in particular. Kim (1989, 1992, 1993) adopts a materialist/physicalist thesis that requires that everything there is be physical. His physicalism is accompanied by two major principles that guide his discussion of mental causation: the principles of the causal closure of the physical domain and of explanatory exclusion. Kim formulates the principle of the causal closure of the physical domain in several fashions, and his use of it seems to reflect weaker and stronger versions.[8] One version is that the sciences tell us what is physical. A stronger version is that all entities and events are to be explained, at least in part, by physical theories. A still stronger version requires that all entities and events are to be explained only by physical theories. And most strongly, these theories will be theories of physics. Since the last two versions seem to build mental epiphenomenalism into Kim's high-level hypotheses, I use the first two weaker versions in order to avoid begging the question and to keep the question an empirical one. Thus, I interpret the principle of the causal closure of the physical domain as a corollary of a scientific materialist hypothesis. The closure principle asserts that what exists and has causal powers is what our best scientific theories in all the various scientific disciplines tell us exist and have causal powers. The principle of the causal closure of the physical domain, then, limits the range of causal explanatory theories by excluding theories from outside the domain of scientific disciplines.

By the principle of explanatory exclusion, Kim means that there can be only one complete and independent causal explanation of physical events. Kim also uses this principle in crucial ways that seem at times to eliminate mental causation. To avoid this consequence, I interpret the principle of explanatory exclusion in a relatively weak fashion as demanding that from the allowed explanatory theories there will be only one complete and independent explanation. Such a complete and independent explanation need not, however, come from just one or two domains of physical explanation, for instance, the theories of physics or those of physics and biochemistry. Thus, I am reading Kim so as to leave the burden of the work of excluding mental causation to supervenience.[9]

Kim argues that mental causation supervenes on physical causation. Briefly, suppose that Pat's thought that eliminativism might be false (m) stirs up fears of rampant supernaturalism (m*). By the doctrine of supervenience, we know that the mental effect is itself determined by its supervenience base, some neuronal state, (p*). So, if Pat's thought causes her fear, then it does so by means of causing the fear's supervenience base. But Pat's thought itself is determined by its own supervenience base, another neuronal state, (p). Then, if we make what Kim believes to be the highly plausible assumption that there are causal relations between p and p*, we can argue that the supervenient causal relation between m and m* is determined by the causal relation between p and p*. Kim contends that the supervenience of mental causal relations on physical causal relations is merely an instance of the general relation of the supervenience of macrocausation on microcausation. Thus, although in one sense the lower-level events do all the causal work, nevertheless, insofar as we attribute causal powers to macroevents and entities generally, the same sort of attribution is appropriate in the case of mental causation. Mental events and causation have not been eliminated, but whatever causal power mental events have is completely derivative and dependent.[10] Applying these contentions to the moral realm gives us ethical epiphenomenalism. What sort of response can the integrationist make to this challenge that her account of moral agency does not really identify genuine causal factors that contribute to an explanation of the acquisition and activity of moral agency?

I take the exclusive supervenience explanation thesis to be an empirical claim to be established on the basis of an examination of cases of higher- and lower-level explanations. But the examination of such cases, especially attempts to explain biological, psychological, and social phenomena completely in terms of underlying physical and chemical phenomena, seems to show the insufficiency of supervenience explanations and so to render the exclusive supervenience explanation thesis problematic (Garfinkel 1981; Kincaid 1987, 1988; Menzies 1988). The basic reasons for the failures of supervenience explanations to be sufficient are that (1) higher-level phenomena may be described functionally and lower-level subvenient phenomena may not be, and (2) lower-level explanations sometimes presuppose information from higher-level accounts.[11] The first sort of case is common in biology, psychology, and the social sciences, where functional kinds abound. For instance, hearts are described as organs that pump blood, but the parts of the heart are often characterized in terms of their constitutive musculature. Various sorts of materials can and do go into making the same kind of functioning organ. Thus, we might expect that laws concerning the types of musculature constitutive of a heart might not be sufficient to answer questions about its function. Even cells and genes are characterized functionally; and, since these functions may be instantiated by a large number of molecular components, it may not be possible to give purely molecular descriptions and explanations of their functions (Rosenberg 1985a,b; Kincaid 1987, 1988, 1990). Moreover, in these cases and others, lower-level explanations of higher-level phenomena may make use of higher-level information. For example, an explanation of a factory's closing (an institutional phenomenon) because of the availability of nonunion workers who will accept very low wages (an apparently lower-level kind) may make use of macro-level information concerning unions, labor laws, and the ability of corporations to move their operations across national boundaries. Moreover, we may be unable to specify such macro-level information in lower-level terms (Kincaid 1986). In all these cases, lower-level theories cannot adequately explain higher-level phenomena. On this basis, I believe that there is good reason to hold that supervenience explanations of moral phenomena by nonmoral phenomena may also be insufficient. They may not answer completely the fully specified questions for which we are seeking an answer.

For example, Kim asks us to suppose that the moral goodness of an act supervenes on the psychological dispositions of sympathy, benevolence, courage, and honesty (Kim 1990). If so, he claims that we can explain the moral goodness of an action of St. Francis, for instance, because it derived from his sympathy and the benevolence of Socrates because it had its source in courage and honesty. But this example does not make Kim's point. The meaning and role of such psychological dispositions in these explanations presuppose their meaning and role in higher-level moral explanations; thus, they are not themselves candidate kinds for higher-level explanations. Explanations of moral actions are a special form of intentional explanation, involving intentions to achieve moral ends or failures to do so. Sympathy, benevolence, courage, and honesty can function in moral explanations if they are moral virtues. However, to conceive them as such presupposes a theory of moral virtues and thus a moral theory. On the other hand, if these characteristics are considered to be purely psychological or biological characteristics, then they are insufficient to explain the goodness of a moral action since the latter requires an intentional explanation in terms of moral ends. That, of course, is not to say that they would have no role in a complete explanation, for instance, by providing a first-level motivational source that can be appropriated by a moral agent. So if, as I have argued, there are biologically or psychologically based moral sentiments, they may provide such motivational sources. Thus, supervenience explanations can retain a role in a complete explanation of moral agency. However, as such, they are insufficient to explain moral action. Such moral explanations of actions in terms of moral prescriptions based on moral values can be interpreted to have the form of a selection theory, as I have discussed in Chapter 2 (Darden and Cain 1989, pp. 106–29). As such, they share a common structure with the theories of natural selection and operant conditioning. Adaptations, understood in a broad, nonexclusively biological sense, are explained as effects of the interactions of selecting environments with bearers of varying factors that are in some respect beneficial or harmful to the bearers. According to these theories, higher-level phenomena in the selecting environment play a causal role in the distribution of the variants, whether genes, organisms, or behavioral repertoires.[12] In the case of moral agency, moral ends are a part of the higher-level, cognitive selective environment that explain, at least in part, moral actions, as well as the presence, for instance, of certain developed psychological capacities like courage or sympathy.

Thus, there are some scientifically based arguments for believing that the exclusive supervenience explanation thesis is not generally true and also some reasons for contending that Kim's application of it to moral explanations is problematic. Moreover, in their commonsense form, explanations of moral and nonmoral actions that invoke such top-down and same-level causality are all too familiar to us. For instance, I might explain why the CEO of a large company neglects the health and safety of his employees by claiming that he is fundamentally an immoral person. Or I might explain a health care worker's tireless efforts on behalf of her patients by saying that she is a morally good person. However, the appeal to such commonsense explanations may appear question begging. Granting that the exclusive supervenience explanation thesis may not be generally true, and granting that one can make up examples from common sense of how supervenience explanations might fail to be sufficient explanations in the moral realm, what *scientific* evidence is there that *in fact* supervenience explanations fail in the *moral* realm?[13] I shall argue that there is enough scientific evidence for both same-level and top-down causality on the part of psychological systems that it is not implausible scientifically to claim that the moral level constitutes a distinct level of causality. If I am correct, the integrationist can use scientifically based reasons to resolve the supervenience dilemma, thereby avoiding ethical epiphenomenalism while keeping her scientific and naturalistic commitments. To do so, I return to the scientific bases for my account of moral agency in the theory of the self-system.

I have based a theory of moral agency on a psychological theory of an agent's capacities, specifically, Bandura's social cognitive theory of agency – his theory of the self-system. How does the theory of the self-system bear on the exclusive supervenience explanation thesis as applied to the moral realm and to ethical epiphenomenalism? We can understand same-level causation to include thoughts causing other thoughts and feelings, and feelings causing other feelings and thoughts. If behavior is understood to be bodily movement and to be at a lower level than the mental, then the mediation of behavior by thoughts and feelings is a form of top-down causation.[14] Given these understandings of same-level and top-down causation, I conclude, on the basis of Bandura's studies, that the self-system, by means of both its self-evaluative and self-efficacy subsystems, is causally efficacious in the moral realm both in a same-level and a top-down fashion. Moreover, there is evidence that the self-efficacy subsystem exercises top-down causality on neurophysiological processes themselves. In a recent set of studies, Bandura and his collaborators have demonstrated that perceived self-efficacy causally mediates the presence of crucial neurotransmitters and hormones operative in actions in fear-inducing situations.[15] This finding, although more controversial, is consistent with (1) animal studies on the role in stressful situations that anticipation and control, or the lack thereof, have on an animal's physiology – recall our discussion of Seligman's studies of learned helplessness in dogs; (2) the accumulating body of evidence that some stress-related illnesses and diseases in humans have psychosocial bases (Kiecolt-Glaser and Glaser 1987); and (3) the mediating role of cognitions in pain control (Bandura, 1986b). There is, then, some scientific evidence for same-level and top-down causation on the part of a psychological system that is intimately connected with moral agency, that is, the self-efficacy system that is responsible for assessing one's capabilities for doing specific tasks, including the achievement of moral ends.

How might Kim respond to this evidence for same-level and top-down mental causality? He argues that laws will be found uniting the neurophysiological events p and p* on which m and m*, for instance, perceived self-efficacy and reduced fear, supervene. In addition, he claims that it will be shown that p is both a necessary and a sufficient condition for m, that is, some neurophysiological event(s) bring about the given state of perceived self-efficacy. Since I take the exclusive supervenience explanation thesis to be an empirical hypothesis, I do not wish to rule out this possibility a priori; however, the evidence in the case of perceived self-efficacy does not point in the direction of the sufficiency of p for m. Rather, p seems to be only partially responsible for the level of perceived self-efficacy. Other factors, including mental processes – for instance, learning by practice, modeling, and instruction – also seem to be required to bring about perceived self-efficacy. So, I take it that the current state of scientific evidence shows that the exclusive supervenience explanation thesis fails in the area of moral agency. This appeal to current *scientific psychological* evidence may not persuade physicalists like Kim who are committed to the strongest version of the principle of the causal closure of the physical domain. However, as I have indicated, such a priori commitments to the necessity and sufficiency of explanations in terms of the causes discovered by physics seem to beg significant empirical questions. Materialists need not be physicalists unless the evidence demands it. I see no indication at the present stage of scientific work that it does. Thus, I conclude that, at present, retentive reductionists like Kim cannot make use of the exclusive supervenience explanation thesis to argue for ethical epiphenomenalism.

In sum, then, it is scientifically plausible to claim that the self-system, through its component evaluative, monitoring, and capacity-assessing subsystems, exercises same-level and top-down causality with respect to moral thought and action. Thus, I conclude that the exclusive supervenience explanation thesis fails. Consequently, the first horn of the supervenience dilemma is blunted and explanatory ethical epiphenomenalism avoided.

Moral agency is genuinely causal. At the same time, an advocate of the integrationist account of moral agency can allow that supervenience explanations play a necessary role in a complete account of moral agency.

Emergent mentalism now emerges as a genuine alternative to both retentive and eliminative reductionism. To understand what I mean by emergent mentalism and its role in the integrationist account, let's consider that notion in more detail. In a helpful discussion of modern notions of emergence, Robert L. Klee (1984) finds that a property, P, has been characterized as emergent with respect to a system's lower-level microstructure when (and possibly only when) either (1) P is unpredictable in principle from the microstructure or (2) P is novel with respect to the microstructure or (3) the microstructure exhibits more variation than does the level of organization on which P occurs or (4) P has a determinative influence on at least some of the properties of the microstructure.[16] The first characterization of emergence as unpredictability-in-principle seems improbable, except in the case of quantum phenomena. I shall not discuss that sort of emergence. Emergence as novelty needs specification. Even if novel properties are properties that are new in kind relative to previously extant properties, unless these properties are also causally efficacious, they remain epiphenomenal. Emergence as relative variation seems to be an alternative way of capturing the phenomenon of multiple instantiation. As such, it entails the concept of levels but is, nevertheless, compatible with epiphenomenalism. It is the fourth sort of emergence – what we might call *strong emergence* – that is a feature of moral agency. In contrast with epiphenomenalism, strong emergence entails not only the existence of distinct levels but also the causal efficacy of higher-level phenomena, both the same-level causality of psychological factors affecting other psychological factors and top-down causality, psychological factors affecting neurophysiological factors.

Moral agency is a strongly emergent phenomenon involving higher-level causal powers but, unlike the emergence envisioned by Lloyd Morgan, it is not an inexplicable phenomenon to be accepted with "natural piety" (Morgan 1923, p. 4). It requires for its origin, maintenance, and operation both lower-level subvenient and higher-level natural and social selective forces. Moreover, there is no reason why such forces and their modes of operation cannot be discovered and understood. Indeed, this seems to be exactly what is happening. Consequently, an integrationist can continue to take seriously the fundamental concepts of emergence and continuity in nature that seem to be central to the materialist and Darwinian perspective that she embraces without either falling into reductionist pits or sailing over mystical cliffs.[17]

Thus far in my argument for an emergent mentalist understanding for our account of moral agency, I have only addressed the retentive reductionist end of the reduction continuum. We need now to examine another argument in support of the reductionist dilemma, this time coming from the eliminativist side of that continuum.

7.5 THE ELIMINATIVIST OPTION AND THE STATUS OF FOLK PSYCHOLOGY

The retentive reductionist anticipates that an adequate theory of moral agency will be part of a final scientific framework because it will be reduced to a completed neuroscientific theory or lie somewhere near the retentive side of the reduction continuum. The eliminativist, on the other hand, contends that the scientific future will reveal that theories of moral agency have been eliminated or that only fragments of such theories will survive and these remains will bear very little resemblance to folk psychology or scientific psychological theories that make use of the intentionalistic categories of folk psychology. The eliminativist argues that if folk psychological or, more generally, intentionalistic categories are required for the description and explanation of moral agency, then we can anticipate the absence of descriptive and explanatory accounts of moral agency in a completed scientific framework. Thus, despite the apparent victory over retentive reductionism, an integrationist proponent of moral agency is still faced with a reductionist predicament, this time induced by eliminativist considerations. To the extent that she wishes to address moral agency, it appears that she must abandon her scientific commitments, and to the extent that she is loyal to her scientific commitments, she has to relinquish her attempts to address moral agency. In order to solve this new problem, I need to lay out in more explicit and coherent fashion the eliminativist critique of folk psychology and those scientific psychological theories that make use of folk psychological categories. My hope is that when these critiques are examined in more detail, we may still pursue the search for moral agency without abandoning our aspirations to make ours a scientific account of moral agency.

Folk psychology is the framework that we use to talk about our mental life and its effects on what we do and say and feel. Besides beliefs and desires, we talk about perceiving and perceptions, imagining and images, thinking and thoughts, wanting and wants, deciding and decisions, fearing and fears, being angry and what we are angry about, hoping and hopes, and so on. The folk psychological conception of mental life presents it as a set of intricately connected states, activities, and their objects, for instance, thinking and thoughts. Philosophers have come to call this combination of states, activities, and their objects *propositional attitudes. Attitude* refers to the mental state or the activity and *propositional* to the object of that state or activity. Moreover, we have to remember that our mental life is much more than propositional attitudes. Philosophers sometimes refer to the non-propositional aspects of our mental life, for example, feelings and sensations, as *qualitative* or *phenomenal* aspects. But, in addition to propositional attitudes and qualitative mental states, we might add those mental processes and states to which we have no direct access, for instance, processes of perception, thinking, bodily regulation, and behavior. Propositional attitudes are distinguished from nonpropositional mental states by the fact that they are *about* something. They are said to take an object. This feature of mental states has been called *intentionality*.[18] Propositional attitudes form complex networks of connections. Our mental states are connected in complicated ways with each other, with various causal inputs and with the actions they bring about. We can take this complex network of connections to constitute the framework of folk psychology. We use the framework of folk psychology to understand, explain, predict, modify, and control both our own and others' behaviors, thoughts, desires, wants, feelings, and so on. In fact, we can and do assess both the cognitive and pragmatic value of folk psychology on the basis of how well it achieves these goals for us. Indeed, it seems to be a very successful framework for accomplishing these aims.

Given the success of folk psychology, why would reductionists want to reduce it to something else and the eliminativist to eliminate it? Such projects seem doomed to failure. Moreover, it seems evident that we have beliefs and desires. In fact, how could anyone believe that she has no beliefs? She would have to have the belief that she had no beliefs in order to have that belief! The eliminativist position not only seems to be empirically incorrect but also conceptually self-annihilating. If eliminativism eliminates anything, it eliminates itself!

We can divide up the eliminativist's critique of folk psychology into two parts: first, its inadequacies as an account of other people's mental capacities and their effects and, second, its ineptitude even for accounting for our own mental life and explaining why and how we do what we do (P. S. Churchland 1986).

Consider first the fact that there is a large range of normal adult human mental life that folk psychology does not tell us very much about, for instance, perception, memory, learning, imagination, and intelligence. There are many phenomena that our brain seems to be involved in that we cannot learn much about from folk psychology, for instance, our biological appetites, sensations, feelings, reflexes, posture, and the ability to move our limbs. Nor does folk psychology help us very much with understanding mental illness or the behavioral and cognitive deficits of people with damaged brains.

A champion of folk psychology can grant all this and still lay claim to an important chunk of intelligent human behavior and mental capacities that center on our linguistic abilities. For it is especially in our linguistic capacities that intentionality is manifested. This is undoubtedly the case, but the eliminativist is not willing to grant this more modest claim to the folk psychologist. She sees several problems with this version of folk psychology. First, intelligent behavior does not seem to be restricted to humans with verbal abilities. Indeed, nonverbal humans – deaf-mutes, for instance, or preverbal infants – exhibit much intelligent behavior. This is also the case for many mammals and living organisms. Second, much of our knowledge seems not to be in sentence form. It has been called *tacit knowledge*. We seem to have many, if not an indefinite number, of beliefs that we have never enunciated to ourselves, that is, that have never been formulated as mental sentences. I not only believe that the greenhouse effect is a recent phenomenon, I also believe that it did not begin before the year 1776. Nor did it begin before the year 1775. Nor 1774. So, if I do not represent these facts as beliefs, where by *belief* I mean a mental sentence, how do I represent them? Finally, it has become clear, with the attempts to make intelligent machines that imitate the accomplishments of humans, that these machines require a large amount of background knowledge to accomplish very simple tasks. Researchers in computer science and artificial intelligence have come to call the problem of trying to understand this background knowledge and attempting to build it into the machine the *frame problem*. We humans seem to learn enough or have enough built-in cognitive capacity to solve the frame problem for a large range of situations. However, it is not clear that the background knowledge that we have and put to use in various frameworks is something that can be accounted for in the folk psychological framework. It does not appear to be merely a complicated set of beliefs and commands formulated in mental sentences. Given these various problems about the cognitive and pragmatic adequacy of folk psychology with respect to other persons, the eliminativist hopes that we will at least realize that folk psychology is, at best, only a partially adequate theoretical framework.

But the eliminativist also contends that folk psychology provides an inadequate account of our individual mental lives. However, at first blush, this contention appears highly problematic. That we each have beliefs and desires and that we seem to be immediately aware of what they are appears to be transparently evident to us, so much so, for instance, that we can neither doubt that we have a specific belief, if we do, nor wonder about its content. Moreover, even though we cannot literally perceive the mental states of others, it seems clear that we can have, by means of introspection, an immediate internal perceptual contact with our own mental life. In addition, we are in a privileged position to know why we believe what we do believe and why we do what we do. Folk psychology as applied to others may be theoretical both in the sense of being about unobservables and in the sense of not providing absolute certainty, but it is a valuable account of each individual's mental life.

Once again, the strategy of the eliminativist is to erode our confidence bit by bit. Take first the claim that we know why we believe and do what we do. This may be true sometimes, but each of us, if we reflect on it, can probably come up with instances in which we were not exactly sure why we did what we did or clear about the reasons for our beliefs. In addition, we can probably recall times when we were not very clear about what we believed on a certain issue or what we desired or wanted in a particular situation. Indeed, it may have been the case that a behavior of ours was the result of a desire that we had explicitly disavowed and a belief that we had explicitly denied. It is only after the fact, in a moment of honesty, that we are willing to admit this to ourselves; or, perhaps, the evidence is overwhelming that this is the case, although we are not willing to admit it to ourselves. For instance, the evidence points strongly to the conclusion that what I did was an act of jealousy, but I refuse to admit it. None of this necessarily involves any abnormality or mental illness. These sorts of cases do not, of course, threaten the entire framework of beliefs and desires. We still can claim that we have beliefs and desires and the other states and processes that make up folk psychology. It's just that we don't always apply these concepts correctly or know exactly to what they apply (Nisbett and Ross 1980, Chs. 9 and 10).

Even with respect to more immediate sorts of cognitive activity where we might think that there is no likelihood of error, we seem to err (P. S. Churchland 1986, p. 308; P. M. Churchland 1988, pp. 76–9). We can mistake a sensation of extreme cold for one of extreme heat. We can dream that we are suffering some sort of ache or pain, only to discover on awakening that it was merely a dream. We can expect to sense or feel something and, because of that, err and think that we do, even when we do not because our expectations were incorrect. Thus, there seems to be good evidence that in fact we do err both about our propositional attitudes and about our qualitative mental life.

Given the fact of these errors, it is easier to see that we have *mistakenly* taken certain aspects of introspective experience to indicate that it either cannot or does not, in fact, suffer from error. And, because of that, we mistakenly believe that folk psychological claims and the framework built from such experiences are not theoretical. One such misleading aspect of introspective experience is its immediacy. We tend to think that because there appears to be no temporal or other mediation of our introspection, there is no source of error. But this does not follow. Perception of the external world also appears to be immediate, but in fact, we do make perceptual mistakes. Immediacy is not a guarantee against error. Moreover, the failure to discern any mediation in our introspective activity does not mean that such mediation is not present. We do not discern the complicated mediation, both cognitive and noncognitive, involved in perception. But there is good reason to believe that many intervening processes are involved in our ability to see, hear, taste, smell, touch, and so on. In addition, it seems clear that cognitive processes similar to inference or interpretation are present in some perceptions. The presence of such processes increases the likelihood of error. Indeed, if we accept what has come to be called the *network theory of meaning,* that is, that the application of any concept involves an elaborate conceptual scheme, then it does not seem likely that even introspective categories will always be applied without error. Sometimes we make use of sophisticated mental concepts in describing our inner states in a noninferential fashion. For instance, I claim that I introspectively perceive the kind of anxiety that comes from a lack of an adequate sense of self-efficacy in writing about such complicated materials. The fact that such sophisticated, introspectively based claims appear to be noninferential and immediate should not, by the fact of their noninferential and immediate appearance, lead us to think that they cannot be in error. It is this possibility of error and thus inadequacy in description, in categorization, and subsequently in explanation, prediction, understanding, and control that makes folk psychology, even when we apply it to ourselves, a theoretical framework. As such, we can conceive of its being supplemented by, revised by, reduced to, or eliminated by a better scientific framework.[19] But, of course, abstract possibility and actual fact are two different things.

But there are two further arguments in behalf of the retention of folk psychology that I need to examine, the first an introspectively based claim for the stability of the folk psychological framework and the second an argument that eliminativism is self-refuting. Consider the apparent fact that if we are aware of something, for instance, a sunset or the pleasurable feeling that the sunset gives us, we are aware that we are aware. Is not this immediate access to our awareness of something, whether internal or external, not only immediately apprehended, but apprehended without application of any conceptual framework and thus without the possibility of erring? Descartes, of course, argued that even if we were in error about what it was that we were perceiving or introspecting, we could not be in error with respect to the fact that we were perceiving, introspecting, or performing some sort of mental activity. The skeptic, including the eliminativist, is not persuaded by this reasoning. She points out that there is good evidence that such processes of awareness are mediated by subcognitive processes that could go awry. In addition, she points to the conceptual nature of the claims made about awareness of either external perceptual activity or mental activity. For her, the presence of the conceptual indicates that there is a possibility of error. But, even if no error were present in such cases, to satisfy those who are not already committed independently of empirical evidence, the absence of error would have to be established empirically, not a priori as a matter of definition. Indeed, the skeptic points to a person suffering from rare neurological problems such as those of blindness denial. Such a person continues to affirm that he is seeing something when, in fact, he is not. What this sort of evidence of error seems to point to is that such a basic feature of the folk psychological framework as awareness or consciousness may need reformulation or even be eliminable.

But that brings us to the argument that eliminativism is self-defeating. We wonder how in the world eliminativism could ever happen. The claim that beliefs, desires, and awareness can be discarded from our understanding of our mental life is itself a belief. We can have knowledge that beliefs, desires, and awareness are eliminable only if we have a justified true belief about that. So, if it is the case that they are eliminable, we can never know it; or if we do know it, then in fact we do not know it because we cannot know anything without having beliefs. But the attempt to eliminate the eliminativist that we have just described begs the question; it presupposes what is under question. It presupposes, for instance, that in order to have knowledge or claim that there are no beliefs and desires, one must have beliefs and desires. It is, in fact, saying that the only way to legitimately discuss our cognitive achievements is to use the folk psychological framework. But that, of course, is exactly what the eliminativist is challenging.

I have described folk psychology as a framework for understanding our mental life, a framework that involves ascribing intentional states and processes to ourselves and others. That tells us why we might call the framework a *psychological* framework, but why *folk?* The idea behind the description of the framework as a folk framework is that it seems to be a commonly shared way of describing and accounting for and managing our mental life. It does not seem to be the result of any special scientific endeavor. Whether it has an evolutionary component or not, it is a common starting point. The interesting thing to notice is that having a folk framework for describing, explaining, and managing is not unique to our mental and behavioral lives. Indeed, there is good evidence that we have or had folk physics, folk chemistry, folk biology, folk botany, and so on. What has happened to these folk endeavors? They seem to have given way to scientific accounts. Aristotelian physics is often considered to be a refinement of folk physics. It incorporates some of the perceptions and accounts of the physical world and its activities that seem to come most readily to hand – for instance, that something needs a force if it is to move or that light travels any distance instantaneously. Newtonian and Einsteinian physics have replaced Aristotelian and folk physics, although there is evidence that for ordinary people who have not studied scientific physics and had a chance to integrate it into their working perspectives, their physical world remains Aristotelian (McClosky 1983). Given that folk disciplines disappear as their scientific cousins are developed, one might argue with some good inductive bases that folk psychology will be no different, that it, too, will be replaced by scientific psychology. There seems to be no good empirical reason for making folk psychology the exception. So, the eliminativist contends that she has good reason to believe that folk psychology will go the way of other folk disciplines and, thus, to doubt the adequacy of folk psychology.

7.6 MORAL AGENCY AND THE REDUCTIONIST PREDICAMENT

Thus, although we may feel some confidence that the current scientific evidence supports our rejection of explanatory epiphenomenalism, and thereby turns back the challenge of retentive reductionism, our eliminativist friends are quick to resume the challenge and argue that since the theory of the self-system is formulated in part, at least, in terms of folk psychological concepts, we can expect its eventual elimination. But we can resist this challenge in several ways. We can admit the connection between Bandura's theory and folk psychology and argue that the empirical and theoretical support that has accrued to the former rubs off on the latter. Alternatively, we can argue that there is no intrinsic connection between Bandura's theory and folk psychology, so that the problems associated with folk psychology do not rub off on Bandura's theory. In either case, the empirical and theoretical support enjoyed by social cognitive theories generally and Bandura's in particular secure their presence on the scientific scene against elimination and, if our previous arguments are correct, against retentive reduction.[20] Moreover, in either case, the integrationist concedes not only that our neurophysiological capacities play a necessary role in the exercise of our cognitive and motivational capacities but also that further knowledge of the structure of the interrelationships of neurophysiological and psychological levels will illuminate our understanding of moral agency and its dependence on our neurophysiological capacities. But suppose we concede that both folk psychology and Bandura's theory, as well as other similar psychological theories formulated in cognitive information processing terms, are on their way to oblivion. As we have seen, such an elimination is not equivalent to the elimination of the psychological level itself, a much more formidable task. Some future cognitive psychological theory, a T_{R*}, remains, and its elimination may prove to be much more formidable than that of folk psychology. If the eliminativist is to be successful, she must show either that the reducing neuroscientific theory can explain all the phenomena that T_{R*} does or that T_{R*} reduces in the classical sense to the lower-level neuroscientific theories. At this point, we seem to be left with speculation about scientific futures. I return one more time to this issue of the relationships of scientific and folk frameworks in Chapter 10.

So far, I have tried to show how results from sociobiology, behavioral science, and cognitive behavioral psychology can help us answer questions of relevance, acquisition, and action about moral agency. I have shown how these results support a four-tiered model of moral agency and help us understand how we can come to acquire that complex capacity and put it to work. Suffice it to say that at this point there is currently a good scientific case for the retention of the causally efficacious, cognitive processes that make up the three top levels of my model of moral agency.

NOTES

1. Hempel and Oppenheim (1948) placed three logical conditions on an adequate explanation: (1) the explanation must be a valid deductive argument; (2) the explanation must contain essentially at least one general law; and (3) the explanans must have empirical content. In addition, there was one empirical condition: (4) the sentences constituting the explanans must be true.

2. As a historical aside, it is interesting to note that Freud abandoned this project. The reasons are disputed among historians and philosophers of science. Some believe he dropped the project because he thought it was premature; others believe that he rejected it because he came to believe that it was an incorrect approach (Grunbaum 1984).

3. Sometimes this is taken to be a case of reductive elimination rather than reductive correction. See the later discussion of reduction as a continuum ranging from retentive reduction to eliminative reduction.

4. Supervenience usually comes in three different forms: weak, strong, and global; it has been discussed in terms of a large number of differing subvenient and supervenient factors, for instance, events, properties, parts and wholes, facts, explana-

tions, and causality. I focus on supervenient explanations and their causal counterparts. Kim (1984a, 1984b, 1987) has been most active in developing the
concept.

5. For doubts, see Kim (1989) and Grimes (1988).

6. The supervenience relation applies not only to the psychophysical levels, but
   also to other pairs of levels. Genes are units of heredity; they are factors in the
   determination of phenotypic characters. They are constituted by and instantiated
   in a set of molecules, DNA. But genes for a kind of trait need not to be identified
   with a particular sort of string of DNA. Rather, the supervenience relation seems
   to apply. If two genes have the same subvenient base, then their supervenient
   properties will be identical, but two genes with the same supervenient properties, i.e., the ability to code for a specific phenotypic characteristic, need not
   have an identical subvenient base. Confer Philip Kitcher (1984) and Howard
   Kincaid (1987, 1988, 1990). In the last article, Kincaid argues for a supervenience relation between cells and their biochemical constituents.

7. I shall assume with Kim that multiple realizability entails supervenience. Confer
   Kim (1992, 1993).

8. Kim prefers framing these principles in terms of events and causes. My formulation is not meant to diminish his realist preferences, which I share.

9. In the view of some, including Kim, this reading may unfavorably affect Kim's
   claims about mental causation. However, as a scientific naturalist, I think that
   the weaker readings of the principles of the causal closure of the physical
   domain and explanatory exclusion are required. The stronger versions of these
   principles seem to beg the question about mental causation. Moreover, they
   seem to be less plausible empirically and more difficult to defend than the
   versions that I have adopted.

10. Kim contrasts real and merely apparent causality, using as an example of the
    latter the motion of a shadow on a screen. We might think that the first position
    of the shadow is causally effective in its coming to a second position. That
    would be incorrect. That is shadow causality, merely apparent. On the other
    hand, macrocausation is real; but it does not have the reality of microcausation
    since it draws all of its causality from the latter (Kim, 1984b). It is difficult to
    determine the exact nature of macrocausation, given Kim's account of its depen-
    dence on microcausation. One is reminded of the medieval philosophical ac-
    counts of the relationship between divine causality and created causality. How-
    ever, Kim does not provide us with a metaphysics of potency and actuality, as
    did the Thomistic Aristotelians, for instance. What is needed, it seems to me, is
    some sort of metaphysical analysis of the nature of causality and its interlevel
    operation.

11. Both Menzies (1988) and Kincaid (1987) offer a third basic reason for the
    failure of explanatory sufficiency of supervenience thesis, namely, that the kind
    terms in lower-level theories may be unable to capture the content of the kind
    terms in higher-level theories. They offer the following example. Suppose that
    the supervenience base determines multiple properties on the higher level. It

may then happen that the supervenience base will not adequately explain higher-level phenomena because it runs together higher-level kinds. For instance, the translucence and conductivity of water both supervene on its molecular structure. Consider two higher-level generalizations: (1) water is translucent and (2) water conducts electricity. Assume that the translucence of water and its conductivity supervene on its molecular structure, H_2O. If that is so, then H_2O is responsible for both the translucence of water and its ability to cause electrical shorts. If we assume that supervenience entails reduction, then we are led to the absurd result that the translucence of water causes its ability to cause shorts. I do not find this example convincing, since the exclusive supervenience explanation thesis does not require reduction. Nor need the same molecular property of H_2O be responsible for both its translucence and its conductivity, even if both of these properties supervene on H_2O.

12. The causal network is even more complex. For instance, in organismic natural selection, the phenotypic traits that are selected for exercise a causal role in an organism's interaction with the selecting environment. This interaction can be either bottom-up, for instance, from the organism level to the group level, or same level, for instance, from one organism to another. In addition, there can be top-down effects of the organism on, for instance, its genetic constituents in subsequent generations. Of course, the genes themselves play a bottom-up causal role in the constitution of the phenotype.

13. Brink (1989) argues for mental causation in ethical actions using both common-sense ethical explanations and a general argument for higher-level causal powers that derives from Garfinkel (1981). Garfinkel's general hypothesis about higher-level causality is confirmed by the kinds of scientific evidence discussed by Kincaid, for instance. I intend my argument to support Brink's and Garfinkel's positions by using well-supported explanations in scientific psychology, specifically social cognitive theory.

14. For an importantly different conception of behavior that nevertheless leaves a role for mental causation, see Dretske (1988).

15. Bandura (1986, 1991) and his colleagues (Bandura et al. 1985) have shown that a high level of perceived self-efficacy is responsible for low levels of blood plasma epinephrine and norepinephrine during interactions with phobic objects such as spiders and that moderate perceived self-efficacy brought about large increases in these catecholamines.

16. Klee argues that the first three characterizations can be understood in such a way as to maintain the thesis of microdeterminism. He contends, mistakenly I believe, that there is no evidence for emergent properties as causally efficacious.

17. Kim (1992, 1993) argues that the emergentist's and nonreductive physicalist's positions are essentially identical with respect to mental causation and that they are both essentially unstable. If they want to maintain their physicalist credentials, they must, in Kim's view, embrace either eliminativism or the sort of reductionism he favors. I have attempted to show that the advocate of an integrationist naturalistic ethics can retain her materialist commitments by embracing

the necessity of supervenience explanations but, at the same time, by rejecting the thesis that supervenience is explanatorily sufficient, retain a stronger sense of emergence than Kim allows.

18. This notion of intentionality should not be confused with the ordinary meaning of the term, i.e., an intentional action as one that is done with knowledge and on purpose. The latter is also intentional in the technical sense I have introduced, but not all mental states that are about something need be actions done with knowledge and purpose.

19. It may be that for both external perception and internal introspection there is a set of nontheoretical categories or concepts in terms of which perceptible things and introspectable mental states are immediately accessed. Such categories may be the result of our genetic endowment. In this case, the folk psychological framework is an evolutionary attainment. But it would still remain to be seen how adequate it is to the realities that it represents. In that sense, it remains theoretically and epistemically replaceable (Rottschaefer 1976, 1978, 1987).

20. Connectionist theories of the mind pose a prominent challenge to scientific psychological theories formulated in terms of information processing categories. These theories model mental processes on parallel processing in networks and can be interpreted as doing without intentional categories (P. S. Churchland 1986; P. M. Churchland 1989; Bechtel and Abrahamsen 1991; P. M. Churchland 1995).

IV

A scientific naturalistic account of moral agency

8

The adequacy of moral beliefs, motivations, and actions

How can biological and psychological explanations serve as justifications?

8.1 FROM ACQUISITION, ACTION, AND RELEVANCE TO ADEQUACY

I have claimed that a satisfactory account of moral agency must answer questions concerning (1) how moral agency is acquired, (2) how it is put into action, (3) whether the sort of agency that is appealed to is morally relevant, both substantively and functionally, and, finally, (4) whether and how that agency is morally adequate or justified. As an integrationist, I am attempting to develop scientifically based answers to these questions, answers that are both well supported scientifically and account for the phenomenon of moral agency. In particular, I have used the findings of biology and psychology to elaborate an account of how we acquire moral agency and put it to work that adequately describes the phenomenon of moral agency and explains both its acquisition and its execution. Even if I have succeeded in this task, a large and very important question remains about whether and how such scientific findings and theories, that is, scientific facts (indeed, any sort of facts), can help us answer questions about whether our moral beliefs, motivations, and actions are morally justified or justifiable. It is this question that I now undertake to answer. I begin in Section 8.2 by laying out in more detail what is usually thought to be involved in the justification of moral agency. We shall see that, according to a traditional philosophical view of justification, answers to questions of acquisition and action do not help us to answer the question of how we justify moral agency. Indeed, the use of such factual material to answer normative questions is considered to be fallacious. In Section 8.3, I turn to naturalized epistemology for a model of justification that will enable us to overcome the barrier between facts and evaluations. In the following sections (8.4–8.6) I apply this model to the justification of moral beliefs, motivations, and actions. Then in Section 8.7 I discuss some general features of the scientific naturalistic justification of moral agency I am proposing. Finally, in Section 8.8, I attempt to turn back the separatist challenge that our efforts are in vain because they run afoul of at least one version of the naturalistic fallacy.

8.2   A TRADITIONAL ACCOUNT OF THE JUSTIFICATION OF MORAL AGENCY

What about moral agency needs justifying?

What sorts of things need moral justification? In general, when we consider an individual, we can ask the question of justification about a person's moral beliefs, motivations, actions, and practices. We could refine and extend this list by considering someone's attitudes, commitments, lifestyle, and so on. Of course, we need not confine ourselves to individuals; the actions, motivations, beliefs, and practices of groups, communities, institutions, societies, and nations can also be considered with respect to justification. My focus shall remain on individuals and their moral beliefs, motivations, and actions. It is common to speak of morally good or bad persons. We apply these assessments both to ourselves and to others. But assessments of practices, actions, motivations, and beliefs seem to be more common.

Consider Anne's speaking to a coworker about a racially prejudicial remark. Assume that her action is a moral action as opposed to an amoral one. Anne can assess each of the aspects of her anticipated action in terms of whether or not it is morally right, that is, whether it achieves what is morally good, and in terms of whether she ought to perform it, that is, whether she has a moral obligation to perform it. Roughly, she can tell if she has a moral obligation in terms of whether in the situation, and because of her own role in the situation, she is required to try to achieve the moral good that the successful performance of her action would bring about. According to this traditional logic of assessment, then, moral rightness and obligation are functions of what is morally good. The general idea is often expressed as a truism: Good is to be achieved and evil avoided. The major point of these considerations is the goal-centered nature of assessment. The achievement of proper goals functions in the judgment about the moral adequacy of both anticipated actions and moral reflection on past actions. We are interested not only in achieving the end of moral knowledge, *true* moral beliefs, but also the ends of morality, what is morally *right* and *good.* So, we are also concerned with morally good motivations and actions, as well as with moral knowledge.

Moral beliefs, motivations, and actions

By thinking in terms of the justification of moral beliefs, motivations, and actions, we are using a philosophical analysis that is fairly close to common sense. We consider an agent to have certain beliefs about what it is right and wrong to do and what is morally good and bad. These beliefs motivate her to perform actions. The agent attempts to discern what is the right thing to do and what she ought to do and tries to do it for the proper motives. But we do not need to stay with this commonsense model of agency; indeed, as integrationists, we have good reason to use the model of moral agency that has emerged in our considerations of the biological and psychological bases for the acquisition and implementation of moral agency. We can consider the function of each of the four levels separately or together – either all levels functioning in unison or some doing so. With respect to at least three levels, there is a cognitive component with cognitive processes and functions whose optimum functioning has to do with the end of ascertaining what is the case. For the behavioral, reflective, and self-reflective levels, the cognitive states are representational, and their optimum functioning is characterized in terms of accuracy of representation. Although base-level states probably are not fully cognitive, they seem to have some representational capacity. However described, the states and processes of the base level also have end states that constitute their optimum functioning; these end states, insofar as they are representational states, may be analogous to accurate representations, the end state of processes that are constitutive of knowledge. Similarly, each of the levels serves a motivational function. Consequently, the representations involved in desire and motivation, which function at each level as action-inducing mechanisms, must also be assessed for their adequacy in moving the agent to the successful accomplishment of a moral action. Finally, the actions themselves, as well as their consequences, can be assessed in terms of their moral correctness. Usually true moral beliefs and properly motivated moral desires and intentions lead to morally right and good actions.

In order to get an idea of how moral agency might be justified, we need to understand the nature of justification. Let's start by focusing on a traditional general account of justification.

What does it mean to justify something?

Suppose that Jill is walking home from school; as she passes by the yard of one of her neighbors, she sees her friend's baby brother fall. No one seems to be around, and Jill sees that little Jimmy has cut himself badly. She immediately runs over, picks him up, and takes him in to his father. Suppose that Jill thinks that she did the morally right thing. She believes that helping people in need is the right thing to do and thinks that she has an obligation to help people in need when she can. On what basis does *Jill* make these judgments? What justifies them? *We* also probably think that responding to Jimmy's need was the right thing to do. On what basis do *we* make these judgments? What justifies them?

These questions about justification are different from the ones that we asked about acquisition, action, and relevance. We are not asking how Jill acquired the cognitive, emotional, and behavioral capacities she needed to discern that Jimmy was in need, to feel and believe that she ought to do something about it, to be motivated to do so, and to know what to do and how to do it. These are all questions about acquisition. Nor are we asking how she put these capacities to work. What moved her to act, how did it do so, and why? These are questions of action. When we have answers to these questions, we can *explain* what she did. But we have not yet answered the question of whether she did what she ought to have done and on what basis she or we make the claim that she did the right thing. Nor have we gotten to the question of justification when we ask whether what Jill did was in the moral realm or not, that is, when we raise the question of relevance. From a commonsense perspective, it seems that questions of justifications are not very difficult to answer. From what we've been told, Jill appears to have done the right thing. We might even think that it's pretty clear that it is right because helping another person in need – when you are able to do so and when it doesn't cause that person more harm than good – is a moral principle that we should follow. So, if Jill acted on that basis, she had good reasons for what she believed and did. She was justified. But what connection, if any, do these *justifying reasons* have with biologically and psychologically based explanations of how Jill learned to act morally and was motivated to do so? In attempting to answer this question, I make use of some helpful parallels from epistemology.

A traditional account of knowledge

A classical definition of knowledge is that knowledge is justified true belief. So, let's take moral knowledge as justified true moral belief.[1] By defining moral knowledge in this way, we are taking a cognitivist approach to ethical knowledge and justification, since we are saying that morality is in part a matter of beliefs that can be either true or false.[2] Naturalist ethicists and scientific naturalist ethicists, including integrationists, need not be cognitivist, but the integrationist view that I shall be arguing for is cognitivist. So, I explore a traditional cognitivist account of justification and knowledge, since it will help us understand the scientific naturalistic approach to knowledge and justification that I shall be advocating. I shall not be excessively demanding in the requirements for what counts for knowledge. I shall not require that in order to have knowledge, one needs to have *certain* justified true belief. That is to say, I shall not demand for knowledge that one be unable to doubt his or her belief. Nor shall I state that knowledge requires incorrigibility. For instance, in order for me to know that I am sitting here in front of my computer, it is not necessary that I cannot be in error. Nor shall I require that in order to know something, to have a justified true belief, it is necessary that I know that I know. In other words, it is not necessary for me to know that I have a justified true belief that I am sitting here in front of my computer. Knowledge does not require knowledge of knowledge. Let us now consider each of the components of our definition of moral knowledge.

We can take beliefs to be mental representations, propositions, or the sentences that we use to enunciate these propositions. Once again, I shall set aside a number of deep philosophical problems about the nature of belief and the ways that philosophers try to explicate the concept or adequately describe the phenomena of belief and believing. Like other sorts of beliefs, moral beliefs can be very broad and general or very specific. For instance, we might hold the very broad – some might say, trivial – belief that good ought to be done and evil avoided. More significantly, but still on the very general level, we might believe that we ought to promote the greatest happiness of the greatest number, as the utilitarians would have us believe. Or we might believe with the Kantians that all persons ought to be treated as ends and never as means. Yet again, we could hold, as some religious people do, that what is morally right is what God wills. Moving from these high-level principles to what we might call *general norms,* we might believe that taking the life of an innocent person is always morally wrong. Or we might hold that telling a lie is never morally permissible. Rather than believing in exceptionless moral norms, we might maintain that in general the taking of the life of an innocent person is morally wrong, but that there may be exceptions, for instance, if there is no choice but to sacrifice one or more innocent persons in order to save many more innocent persons. Supposedly, this was the moral belief used to justify the dropping of atomic bombs on Hiroshima and Nagasaki. Besides moral principles and moral norms, whether absolute or general, there are, on the other end of the continuum, singular moral beliefs, for example, our belief that Jill ought to help Jimmy. This is a belief about a single individual and what that individual ought to do. Between the singular and the general come particular moral beliefs. For instance, the rich have an obligation to pay a higher percentage of their income in taxes than do the poor. Here we are referring to a particular group of people. Building on an analogy with justification for claims for nonmoral knowledge, we might expect that these different sorts of moral beliefs might require different sorts of justifications.

The second key term in the definition of knowledge is *true*. Philosophers have offered many different definitions of this term and have elaborated various theories of truth. There are three prominent theories of truth: correspondence, coherence, and pragmatic. I focus on the correspondence theory. As its name indicates, the correspondence theory proposes that a true belief or proposition is one that corresponds with reality or with what is the case or with the facts. The major variations in this theory have to do with the nature of the correspondence. We can think of correspondence in a simple-minded way as picturing or in very sophisticated ways in terms of complex transformational functions that capture various sorts of isomorphisms, for instance, the functions for the transformation of coordinate systems. Fortunately, we do not need to go into these details for our purposes here. It is enough to think of corresponding as achieving an appropriate fit between representation and reality, where the "fitting relation" depends on both the type of representation and the type of reality in question.[3]

How does the notion of justification fit into the definition of knowledge as justified true belief? One helpful way to think about justification is to consider it as the means for achieving the goal of knowledge, that is, true belief. Or, more traditionally, we can think of justifiers as good reasons for holding that a belief is true. They provide the legitimate grounds for holding a belief. Philosophers have traditionally divided justified beliefs into two types: basic and nonbasic. Basic justified beliefs are propositions that do not need another proposition to justify them. They are sometimes said to be self-evident or self-justifying. Nonbasic beliefs are justified by means of still other beliefs or propositions; for instance, the conclusion of an argument is not self-justified, but is justified by its premises. These, in turn, could be basic beliefs, that is, self-justified, or they may be nonbasic beliefs, requiring still other true propositions for their justification. The picture of justified belief that most readily comes to mind, on the basis of this account of justifiers, is one of a cognitive structure, with the basic, self-justifying beliefs serving as the foundation of the cognitive enterprise and the nonbasic, derivatively supported beliefs constituting the rest of the building. This foundationalist picture of justification is often associated with a correspondence theory of truth because the basic beliefs are thought to be, for instance, perceptual or experiential beliefs that correspond with perceptual or experiential reality. But we should note that, in the correspondence theorist's view, these are not the only beliefs that correspond with reality. For a correspondence theorist, any *true* belief, whether basic or nonbasic, corresponds with reality. The foundationalist picture of justification also denies that there is an infinite regress of justified true beliefs, all of which are supported by nonbasic justifiers. At some point the regress must end. The cognitive building has to have a foundation. Nor will it do, in the correspondence perspective, to have a picture of justification in which some basic or nonbasic beliefs circle back and gain support from other nonbasic beliefs for which the former provide support. Suppose that Jill concludes that she has an obligation to help Jimmy, who is in pain. We ask her what justification she has for her belief. She responds that anyone who is present and able has an obligation to help someone in pain. We then ask her for a justification of her general moral norm. If she responds that her justification is that the general norm follows from the claim that Jimmy's pain ought to be alleviated, then she seems to be arguing in a circle. Foundationalists have a phobia for such justificational circles and try to avoid them, since they believe that circular justifications do not provide any justification.

In contrast to the foundationalist picture of justification, there is what we might call the *network theory* of justification. The network theory does not require basic beliefs in the sense demanded by the foundationalists, that is, propositions that are self-evident or self-justifying. Network theorists allow that all propositions can support each other in various ways. So, they hold that all beliefs are nonbasic. Network theorists, then, as we can see, allow for some circularity in belief support and do not demand a grounding for knowledge in the same way as do the foundationalists. There seems to be a natural connection between a network theory of justification and a coherence theory of truth. Although we could fill in a lot of details to round out the foundationalist and network theories of justification, we have enough for the purpose of understanding the traditional picture of the justification of moral beliefs.

A traditional account of the justification of moral beliefs

Starting with Jill's belief that she ought to help Jimmy, let's take each of the different kinds of moral beliefs that we have mentioned and consider how they might be justified. We'll use a foundationalist approach, not because it's to be preferred, but because it's probably easier to see how it works. Jill's moral belief is a singular belief in that it is about herself and what she ought to do in a singular circumstance, namely, that of Jimmy's falling and hurting himself. This sort of belief might be considered to be a basic moral belief, that is, one that is self-justifying or self-evident. What we mean by saying that such a belief is self-justifying is – negatively – that it does not need to have other moral beliefs to support its truth. Positively, we mean that it provides its own reason for claiming that it is true. So, we say that it is self-evident or self-justifying. The evidence for its truth comes from itself or from its experiential sources, not from other propositions or beliefs. At this point, I am not arguing that, in fact, Jill's belief is a basic one and so self-justifying or self-evident. I am merely trying to illustrate what it might mean to say that it is so. Now let's move to the other end of the spectrum, to moral principles. Let's take the principle that pleasure is a moral good and that pain is a moral evil. The hedonist makes this his ultimate ethical principle on the basis of which he can judge the moral rightness or wrongness of actions, assuming that he holds that pursuing moral good is morally right and pursuing moral evil is morally wrong. On what basis might the hedonist justify his ultimate ethical principle? At least two ways come to mind immediately. We might think of the principle as a kind of high-level ethical theory that is well confirmed by many experiential or empirical contacts with pleasure and pain. As a high-level theory, the hedonistic principle is not self-justified; rather, it is a nonbasic moral belief. That is to say, it receives its justification from other moral beliefs, in this case, from basic beliefs originating from many particular experiences of pleasure and pain. Typically, however, such high-level moral principles are considered to be generalizations or intuitive inductions for which a few or even one experientially based belief is sufficient to provide justification. The connection between general principle and basic belief that functions as a justifier appears to be much closer in this typical understanding than when the general principle is considered to be a high-level theory or hypothesis. Nevertheless, the justification is derivative and not basic as long as the hedonist is not claiming that his ultimate principle is self-justified or self-evident. Finally, consider the moral norm that torture is morally wrong. How is such a norm justified? The most plausible answer to this question is that it is justified on the basis of other moral beliefs. The hedonist might claim that he can provide us with a sound argument using his hedonistic principle and some other true moral premises to justify this moral norm about torture. On the other hand, he may consider the moral norm to be something like a hypothesis that is confirmed by supporting singular and particular moral beliefs. In both understandings, we say that the moral norm is justified and so a true moral belief because it is supported by other true moral beliefs.

So, if moral knowledge is justified true beliefs about morality, then the foundationalist or the network theorist can claim moral knowledge, provided that she has justification for her true moral beliefs. Of course, since we are not requiring justifiers to be perfect instruments for finding true beliefs, it may be the case that some of her justified beliefs are false. Nevertheless, we can now conclude that it is justified moral beliefs or moral knowledge that would enable Jill, or ourselves, to assert with justification that she has an obligation to help Jimmy. On this basis, she could then respond to questions that she or others might raise after the fact about whether she did the right thing or not.

8.3 A SCIENTIFIC NATURALISTIC ACCOUNT OF THE JUSTIFICATION OF MORAL AGENCY

The problem of relating causes and reasons

Hopefully, we now have a sense of what justification is on a standard philosophical account of it. Now let's return to our original question of whether the knowledge that we have about the acquisition and activation of various facets of moral agency is helpful for answering questions about the justification of moral beliefs. More specifically, can an integrationist who has knowledge of the biological and psychological factors that describe and explain the acquisition and activation of moral agency use that explanatory knowledge to justify her moral beliefs?

In keeping with our strategy of using epistemology to help us clarify issues in ethics, let's consider an analogous problem in the justification of a factual belief. Suppose you are writing a paper on Dostoevsky's *The Brothers Karamazov.* One of your friends asks you how you are doing on it. Your reply is that you are doing just fine. You expect that you'll finish it with relatively no problem. Your friend is the Socratic type, so she asks for the basis of your optimistic assessment of your completion time. You reply that you dreamed last night that you would finish the paper in a blaze of glory. Or you say to her that you heard it last night on TV. Your replies sound like jokes, not justifications of your claim that you will successfully complete your paper. At best, they tell her how you might have gotten the idea that you will have no problem in finishing an excellent paper. But what you need for a justification is something like the following reasons: (1) you have successfully completed similar papers in the past, (2) the Dostoevsky paper is like those you have done, and (3) Dostoevsky hasn't driven you mad yet. The point is straightforward. There is a big difference between where you get an idea about something and what justifies it. You yourself have probably gotten a lot of ideas in various odd or not so odd ways. But the fact that you got the idea doesn't mean that it is correct. Justification of a belief is different from the discovery or origin of a belief. One way to put that difference is to say that the former is concerned with the reasons that are asserted in support of a belief and the latter is interested in the causes of a belief, that is, how you got the belief.

The same distinction seems to be applicable to moral belief. Suppose, for instance, that we grant to the Skinnerian that there can be a science of values in the sense of a scientific discipline that (1) identifies values for the species, a society, or a culture; (2) discovers empirical laws that show how values are related to behaviors and to the conditions in which various behaviors occur; and (3) develops theories that explain these laws by laying out the causal factors that are at work in producing the regularities captured in the empirical laws, and so gives us an understanding of the foundations of values. Nonetheless, such a descriptive and explanatory account of values does not seem sufficient for matters of justification. What we need to know is not just what our values *are* but what they *ought to be.* We need ethics, the philosophical discipline concerned with the study of moral values and obligations, what is morally right and wrong, and the prescriptions that bind us morally, as well as the ways in which we can justify these prescriptions and give an adequate foundation for values. To put it very simply, science deals with facts and causes, but ethics deals with values and justificatory reasons. It may be a fact that food is reinforcing and is a value. And Skinner's science of values, for instance, may help us understand why that is so. But the science of values can do nothing to tell us what we ought to eat or even *that* we ought to eat or that everyone ought to have sufficient food. Nor can it give us any justificatory reasons for what we ought to do. Skinner has confused his science of values with ethics, thus committing the naturalistic fallacy. That is, he has identified what is morally good with what is and has derived what ought to be the case from what is the case. The integrationist errs, so it might be claimed, because she has confused questions about the *causes* of a belief and the *reasons* for a belief. What makes the naturalistic fallacy a fallacy is precisely this sort of confusion between explanation and justification.[4]

A scientific naturalistic model for the justification of moral beliefs

Recall the case of your completing a paper on *The Brothers Karamazov.* Using the traditional account of justification, we have distinguished between how you got the idea or belief that you would successfully complete it and the reasons that may justify your claim. The former refer to the causes for your belief and the latter to the reasons for holding it to be true. On the traditional view, the latter are the justifiers and are distinct from the causal processes, whatever they might have been, that gave you the idea in the first place. But now consider another case. You're out with some friends for the evening. While you are eating at your favorite restaurant, you spot Al and Lisa in the ticket line at the theater across the street. You remark, "There's

Al and Lisa." You're surprised because you thought that they were going to be working on their Dostoevsky papers. Your Socratic friend is again at your side with the familiar question: How do you know that it is Al and Lisa? You say that you saw them and tell her to take a look herself. The area is well lighted. You are at a window table. And you have just had your eyes checked, and they're in good shape. Reflect on how you got the idea that Al and Lisa were in the ticket line. You saw them. But how do you attempt to justify your idea? By your visual observations. In this case, you are using the same means that you used to get your idea in order to justify it. Is that cheating? No! Sometimes the way that we get an idea is also a reliable means for justifying it. That's often the case with perceptual beliefs. So, even though we might want to say that the mechanisms by which we acquire ideas are not always satisfactory for justifying them, sometimes they are. What we want to find are the mechanisms that reliably generate true beliefs. If we do, then we can use them to justify the beliefs that they also generated. If this is right, then sometimes reasons are not opposed to causes since sometimes causes function as reasons. To put it in other terms, the causes that serve as justifying reasons are the ones that are the reliable mechanisms for attaining cognitive ends.

Those who think that there is a naturalistic fallacy can be thought of as accusing the integrationists, Skinner, for example, of confusing how we get to our moral positions with how we justify them. We can grant to the skeptic that not any and every way to achieve a moral stance will also justify that moral stance. But that does not mean that none will. If we can find the mechanisms that in some context or class of contexts generate adequate moral stances, then we can appeal to them in our justifications as well as in our causal explanations. Just as you appealed to your perceptual powers to justify your claim that Al and Lisa were in the theater ticket line, so you might be able to appeal to the reinforcing effects of something to justify it as a value if the process of reinforcement is a reliable process for the identification of what is morally valuable. Of course, nobody is infallible. Maybe you mistook Ivan and Katie for Al and Lisa. And you might mistake that delicious dish of ice cream you were eating when you spotted Al and Lisa for something that is a long-term value for you. The Skinnerian, then, claims that by identifying the laws of operant conditioning, she has identified some of the reliable mechanisms for acquiring genuine values. If she is right, these mechanisms can be appealed to in justifying claims about what is right and what one ought to do. Thus, she could claim that her science of values makes the necessary connection with ethics and ethical justifications.

Let's summarize our discussion up to this point. We have seen that traditionally, justifications in both moral and nonmoral matters are thought of as having to do with providing adequate or good reasons for believing something is so (or such and such) in the case of nonmoral beliefs or for believing that something is morally right, obligatory, or good in the case of moral beliefs. These justifying reasons are distinguished from the causes by which beliefs are acquired. But, according to the scientific naturalistic account of justification that I am proposing, some causes can function as justifying reasons. In the naturalistic model of justification of moral beliefs, moral truth is considered to be the end or goal of moral belief formation. Some of the processes by which moral beliefs are formed and acquired are due to cognitive mechanisms and processes that reliably achieve that goal. These mechanisms and processes can serve as justifiers of moral beliefs. They provide good reasons for holding the belief.

Extending the model to the justification of motives and actions

The key idea deriving from our scientific naturalistic epistemological model is that of reliable mechanisms for attaining cognitive ends. We can think of the question of adequacy as concerned with the optimum functioning of moral agents, either in relation to each other or in terms of the subsystems of which they are composed. The optimum functioning of a moral agent requires accurate representations of the moral environment, proper motivation toward moral ends, and the successful achievement of these ends. On the commonsense application, the requirement is for justified moral beliefs, properly motivated moral actions, and successful moral actions. The good or adequate reasons, the criteria in terms of which we make a judgment about attaining the well-functioning state, are conceived of in our naturalistic model as reliable mechanisms for achieving that end state. Things become more complicated when we apply our idea of justifiers as reliable mechanisms to our account of moral agency. I conceive of each of the levels of moral agency as having reliable mechanisms for producing relatively accurate representations, proper motivation, and successful execution. Thus, on the base level, evolutionarily based moral capacities and learned behavioral capacities function as reliable mechanisms for attaining whatever "representations" and motivations are needed to achieve morally relevant actions. The behavioral level approximates part of the commonsense or folk psychological application discussed earlier. Reliable cognitive mechanisms lead to more or less accurate moral beliefs, and reliable motivational mechanisms bring about more or less proper bases for moral actions, and both together lead reliably to more or less successful execution of a moral action. The reflective level adds the processes that reliably lead to relatively sound moral rules or principles as cognitive and motivational bases for accurate execution. Finally, the self-referentially reflective level brings in processes that reliably lead to conceiving of the self as a moral self. The conception of the self as a moral self, along with self-efficacy judgments, is productive of both accurate reflective and behavioral-level moral beliefs and proper moral motivations, all leading to successful moral execution.[5]

In order to get an idea of how moral agency might be justified, let's work out the structure of justification for each level in a little more detail, using as examples mechanisms from the different levels to illustrate the justification of cognitive, motivational, or actional aspects of moral agency.

8.4 EMPATHIC DISTRESS AND JUSTIFICATION

Recall that we found that there is evidence to think that empathy might be a base and a behavioral-level motivational factor in our moral behavior. Using Hoffman's (1982, 1984a,b, 1988) account, let's see how empathic distress can function as a justifier, that is, be a morally proper motivation. Hoffman's view of what makes for substantive and functional morality fits into the picture that we have thus far developed. He tells us that what makes an act moral is that it is prompted by a disposition either to act on behalf of a person or a group or to behave in accord with a moral norm or standard bearing on human welfare or justice. Moral reasoning may, but need not, be involved in the performance of the action. Thus we see that Hoffman has in mind a paradigm type of substantive moral activity: helping another in need. The functional levels of morality involved are those at the base, behavioral and reflective levels. Hoffman argues that when we examine moral encounters we find that what many of them have in common that is psychologically important is that they all involve potential victims. Why is it that we are concerned with victims? In Hoffman's view, it is because of our empathic dispositions. As we have seen, Hoffman contends that empathic distress has a biological basis and undergoes a normal developmental growth in individuals. Moreover, empathy is amenable to cognitive influences or, to put it into our own functional schema, empathy at the base level can be influenced by our behavioral, reflective, and self-reflective cognitive and motivational capacities.[6] The endpoint of the functional and substantive development through which the newborn, infant, child, and adolescent progress is the capacity to respond to another's need with sympathetic distress that often leads to helping activity.

Granting that all this is factually so, what relevance does it have to the issue of the justification of morality, for instance, the justification of moral motivations? The justification question for moral motivations is this: Were your motivations for your action morally right, and if so, why? If empathic distress is a justifier, that is, if it provides a good answer to that question, then you can respond that since your action was motivated by empathic distress for the person in need, for instance, someone who had been hurt in an accident, your motivations are morally right. They are morally right because empathy is a reliable means for attaining a morally proper motivation, in this case, the motivation to alleviate the injury of an accident victim.

Think of the justification question in another way, recalling our means–ends analysis of the justification situation. The adequacy of the means is judged in terms of its reliability in meeting a proper end. The proper end here is an action that is successful in achieving a moral good. Does empathy usually lead to that sort of action? If so, then empathy can serve as a justifier of your action; it is a reliable means of performing a morally right action. But Hoffman and others have provided empirical evidence that empathy is that sort of motivator. So you have reason to believe that your motivations were ones that lead to a morally right action. You are justified in making the claim that you – or someone similarly motivated in a similar situation – had morally right motivations and performed a morally right action. Of course, one might also argue from her own experience that empathic distress is a reliable means for achieving a morally right action of helping someone in need and a morally right motivator. Thus, Hoffman's theory of empathy as a reliable motivational mechanism supports a folk psychologically conceptualized experience of the positive role that empathy can play in motivating moral behavior and seeing to the successful execution of a morally right action.

But empathy, Hoffman tells us, has its limitations. Since it is based on a bystander model, and since there is a tendency to respond more empathically to those who are present than absent and more to those who are like us than not, it may well happen that empathic motivations fail in certain situations to be reliable motivators. They may fail to move us to help when we ought or to help the right persons when we ought. Or they may move us to help when we ought not. But these limitations on empathy should not surprise us. Given the evolutionary and learning history origins of empathy, we should not expect it to be a reliable mechanism for motivating moral actions in *all* circumstances. Our perceptual powers have their limitations, even though in normal circumstances and with respect to middle-sized objects they are generally reliable. That is, they work fairly well in conditions for which they have been designed to work, designed by nature and by the contingencies of our natural and social learning environments. With regard to nonmoral truths, we need to supplement our perceptual mechanisms in situations, for instance, where the truth we seek concerns nonobservable realities. Similarly, in situations where the stimuli that arouse empathy are not present or when those stimuli are so overpowering that they provoke motivations that do not lead to the best moral action for the situation, we can expect that our empathic motivations need to be complemented by reliable motivational mechanisms deriving from higher levels, for instance, from the reflective level. Empathic distress is an example of a kind of process working at the base (and, perhaps behavioral) level of moral agency as a reliable means for achieving proper moral motivation and correct moral action. Let's turn now to mechanisms of the behavioral and reflective levels.

8.5 SOCIAL COGNITIVE LEARNING MECHANISMS AND JUSTIFICATION

I have assumed that a scientific naturalistic perspective ought to be taken toward traditional justifiers so that they are conceived of as causal processes instantiated by reliable mechanisms. The question then arises about whether the integrationist can give a plausible scientific naturalistic justification of the various kinds of moral beliefs, motivations, and actions. If we are concerned with moral beliefs about what an individual should do in a particular situation or what are justified moral norms or principles, we can conclude that the relevant levels of moral agency that we should be concerned with are the behavioral and reflective levels of moral agency. So, let's explore what sorts of scientific findings we can bring to bear on the issue of the justification of moral beliefs, motivations, and actions associated with these levels of moral agency in our model.

I have used a cognitive behavioral and social cognitive account of the mechanisms of acquisition and action in describing and explaining these levels of moral agency. In particular, I have used Bandura's social cognitive learning theory of agency. Bandura, as well as other social cognitive theorists, has established that three general types of learning processes are involved in successful action: (1) practice, (2) modeling, and (3) symbolic learning. Of course, each of these is itself a complex process involving many sorts of processes at different levels of operation. Take, for instance, Bandura's analysis of modeling. Learning by modeling has, according to Bandura's analysis, four subprocesses: (a) attentional, (b) retentional, (c) motor reproduction, and (d) motivational. Each of these subprocesses, in turn, has component processes. Not every detail of Bandura's account of

201

modeling is confirmed by any means. But Bandura and other social cognitive psychologists have established that modeling is an effective means of learning certain cognitive and motivational processes on the basis of which a person comes to act in a certain fashion that she has seen modeled. Of course, to say that modeling is effective implies neither that it works in all circumstances nor that it works regardless of the condition of the agent or the quality of the modeling. Applying these results to moral agency, we infer that learning through modeling functions as a mechanism for the acquisition of moral beliefs, motivations, and actions. When it functions reliably, it provides justifiers for moral beliefs, motivations, and actions. Similarly, if social cognitive theories are, as I have argued, well supported, then practice and instruction also serve as generally reliable mechanisms for attaining true moral beliefs at the behavioral and reflective levels and adequate moral motivations and correct ethical actions at the behavioral level. Moreover, as with empathic distress, these scientifically identified and described reliable mechanisms are also accessible through our ordinary cognitive capacities. Thus, my scientific naturalistic account of reliable mechanisms builds on a folk psychological one. Now that we have examined mechanisms at the first three levels of moral agency, it is time to take a look at the self-referentially reflective level. Are there any reliable mechanisms at that level?

8.6   THE SELF-SYSTEM AND JUSTIFICATION

In my model of moral agency, the highest level of functioning is concerned with a person's ideal of herself as a moral person and her sense of self-efficacy with respect to what she can be as a moral person. These functions derive from what Bandura has called the *self-system.* If Bandura is correct, persons are capable of developing and deploying the self-system as a reliable mechanism of agency. Applying this idea to moral agency, persons are capable of becoming moral selves. Recall that the self-system has four major informational and motivational capacities and processes that play a central role in explaining human behavior. These concern (1) the formulation and application of individual goals and standards of performance, (2) self-monitoring, (3) self-reactive influences, and (4) self-efficacy judgments. These components function to bring about successful performance through a set of cognitive processes that are self-referential. If the person conceives of herself as a moral self, then the self-system functions in the successful performance of moral behaviors. It does so via the reflective level by setting standards, monitoring performance, and providing self-reactive influences, and on the self-referential level it functions to enable her to view herself as a moral agent. It thereby works to achieve the invocation of moral standards – in preference to alternative standards, for instance prudential ones – to guide one's thoughts and intentions and consequently one's actions. It may also function in terms of choosing among various sorts of moral standards.[7] The self-efficacy system, on the other hand, functions to specify the moral capacities of the person in the sense that it influences not only the person's view of what she is capable of as a moral agent but also the extent and range of her actual performance. In other words, a person's conception of herself and her capacities as a moral agent can play a significant role in guiding reflective level moral beliefs and desires, and these, in turn, can influence behavioral and base-level moral capacities, as well as actions based on the various levels of agency. Thus, the self-system and its components can provide reliable mechanisms for the achievement of true moral beliefs about oneself as a moral agent with respect to one's moral capacities, aspirations, and standards, and it can function as a source of proper moral motivations and actions.

Consequently, we can invoke the processes of the moral self-system as justifiers of moral beliefs, actions, and motivations. However, as with the other levels of moral agency, it is not necessary to contend that these reliable mechanisms for achieving true moral belief, proper moral motivation, and successful moral action are accessible only through the study of scientific psychology. Ordinary experience and reflection might also be used to invoke folk psychological understandings of the various components of the self-system. Ordinary experience and reflection can at least partially identify and characterize the mechanisms that are more fully elaborated in social cognitive theory. From this perspective, ordinary insights and perceptions, as well as the findings of biology and psychology, provide understanding of how the self-system and the other levels of agency work, how they are acquired, maintained, changed, fostered, and put to work in achieving their ends. They too are sources for finding out about ends and about reliable mechanisms for achieving those ends, and thus can also be invoked as means to identify justifiers.

Having seen some examples of reliable mechanisms for achieving the ends of true moral belief, proper moral motivation, and successful moral action at each level of moral agency, we are now ready to explore some of the general features of my scientific naturalistic approach to questions of justification and to respond to some of the general problems and questions such an approach raises.

8.7 SOME GENERAL FEATURES OF SCIENTIFIC
NATURALISTIC JUSTIFICATIONS OF MORAL AGENCY

Identifying moral ends

The key idea in my account of naturalistic justifications of moral actions, motivations, and beliefs is that moral justifications make appeals to identifiably reliable mechanisms for attaining moral ends. An adequate theory of the biologically and psychologically based means by which we attain moral ends enables us to identify mechanisms of moral belief and motivation formation and of moral action. It is to these mechanisms that I appeal in justifying our moral beliefs, motivations, and actions. Thus, my account presupposes that I have been able to identify the moral ends that our biologically and psychologically based capacities enable us to attain. Alternatively, it allows us to figure out that certain psychological capacities are *for* attaining moral ends. How do we identify such ends?

We have already seen, in connection with our discussion of substantive moral relevance, that the question about the nature of the moral good is far from settled. So, in attempting to sketch out an answer to the question of how we identify moral ends, it is necessary to adopt an approach similar to the one that we decided on in trying to determine substantive moral relevance. We recall that the strategy that was adopted was to consider paradigm cases of morally good actions, ones that were least controversial and did not beg questions concerning major differences about what constitutes the morally good. Besides adopting that policy once again, it will be helpful to keep several further points in mind. Focusing on moral beliefs and moral belief formation, I have indicated that there is a range of moral beliefs from moral principles, through absolute moral norms, moral norms that hold generally but not always, particular moral claims, and individual moral claims. In each case, there are various belief-forming mechanisms and various ways of identifying the various moral goods referred to in these beliefs. Second, in identifying a moral good, we can distinguish between intrinsic values, those things that are morally valuable in themselves, and instrumental values, those things, whether morally valuable in themselves or not, that lead to intrinsic values. Having a person's basic needs for such things as food, shelter, clothing, and companionship fulfilled may be either intrinsically valuable or instrumentally valuable or both. In either case, it is a genuine moral value. In the case in which it is instrumental, the adequacy of the judgment about its moral value depends on whether we have discerned accurately the connection between the fulfillment of a person's needs and some other end that is itself either instrumentally or intrinsically moral. In the case in which the value is intrinsic, the adequacy of the judgment about its intrinsic moral value is a function of the moral goodness of the end itself, although it is not necessary that its intrinsic value be open to perceptual attainment. Moreover, there is good reason to think that we need to be value pluralists and to claim that there are a number of intrinsically valuable ends. So, the fact that we have discerned a genuine moral value and have a reliable mechanism to motivate us to achieve that end does not mean that we ought to act to achieve that end. The ethical decision-making process is often much more complicated than that. However, at this point we are trying to figure out a single line of justification with a given reliable motivational mechanism. Our question is not how we come to a justified moral judgment about a particular matter, but how we successfully identify the moral ends that are the goals of our ethical actions. Finally, it is clear from an ordinary, everyday perspective and from a scientific perspective that we have capacities for figuring out what things are for. Our question concerns whether we have such capacities when it comes to moral beliefs, motivations, and actions. Do we have capacities for reliably identifying what is morally good, right, or obligatory?

We can begin by reflecting on our ordinary, everyday capacities for identifying moral ends. We have a stock of moral beliefs about moral ends and justifiers coming from our own perceptions, memory, reflections, inferences, and deductive applications of general norms as well as from authorities such as parents, adults, peers, society, and religion. It is moral beliefs deriving from these sorts of sources with which we begin and that we modify as our moral knowledge grows and is refined both individually and socially. Moreover, moral philosophers of many stripes often refer to moral intuitions and the application of moral norms to particular cases. In addition, they often refer to the process of reflective equilibrium, in which, through the consideration of various sorts of cases, moral intuitions are brought into balance with moral principles, and vice versa, by a mutual adjustment of each. Thus, there seems to be agreement that we have the capacity to discern moral ends, although disagreement about the specifics of how we do so and the nature of these moral ends.

Although as an integrationist I reject an exclusive reliance on the nonempirical and a priori methods of the traditional moral philosophers, and contend that the methods of both traditional moral philosophers and ordinary, everyday moral agents need to be enhanced through the use of scientific methods, I build my case for the ability to identify and discern moral ends on the agreement about the ability to discern moral ends. For instance, as the circle of altruism is extended, we achieve the abilities that reliably lead us to true moral beliefs, properly motivated moral actions, and correct moral actions with respect to larger circles of people. Let us assume that our ordinary claims about discerning moral ends constitute our starting point and that we can make some reasonable, commonsense hypotheses about other moral ends. We can test the adequacy of both theoretical and empirical claims about ends using both our ordinary reflective capacities and our acquired scientific capacities. In other words, we assume that the moral good, just like truth, is not something that is completely apparent to us. Rather, it is something that we come to find out about by ordinary and scientific investigations. So, we start with some reasonably justified, empirically based claims about moral ends and attempt to extend our knowledge of moral ends by achieving reasonably grounded hypotheses about them. Having identified moral ends, we search for reliable mechanisms for achieving these ends. And we appeal to the latter to justify the actions, motivations, and beliefs employed in seeking these ends.

To provide a more specific account of how we identify moral ends, we can distinguish relatively direct or indirect means for doing so. The former are more empirical and the latter more theoretical. Neither approach guarantees that we will always be on the mark with respect to ascertaining and following a genuinely moral end. Indeed, we should not expect the same set of ends to be moral in all contexts. Nor is either approach alone completely satisfactory; but a combination of the two is promising.

What might serve as an empirically based source for justified beliefs about moral goods? Two obvious candidates, in light of the account of moral agency that we have been developing, are empathy and empathic distress. Empathy and empathic distress provide relatively direct acquaintance with a morally valuable end. Moral beliefs mediated by these emotions, although perhaps less direct than the emotional response itself, are analogous to perceptual beliefs and provide empirical access to a moral good. In either case, since the integrationist position I am pursuing is a realist one, I want eventually to ground cognitive attainment of genuine moral values in the world by means of the causal interaction between moral agent and the moral properties of events and things in the material world.

Indeed, it is not uncommon for ordinary moral agents to appeal to perception as justification for moral beliefs or actions. Thus, for example, Jane screams at John in exasperation, "Can't you see that you ought to help him?" As we shall see in more detail in Chapter 9, naturalistic moral philosophers like David Brink (1989) and Nicholas Sturgeon (1985) argue that perceptually based moral beliefs and claims have explanatory and justificatory roles. Moreover, the development of conscience in children shows that different modes of parental moral training affect the type of conscience their children develop (Hoffman 1970, 1988; Radke-Yarrow et al. 1983). The achievement of moral internalization, the ability to act on the basis of what is believed to be morally right, especially in situations of conflict with one's self-interest, depends markedly on the childrearing techniques used by parents. Hoffman (1970, 1988) distinguishes three types of childrearing techniques: power assertive, love withdrawal, and inductive. Only the last shows correlation with moral internalization, and Hoffman contends that there is good reason to think that the relationship is a causal one. Through inductive techniques, parents point out to their children the beneficial and harmful effects of their actions on others in a manner appropriate to their ages. The use of power assertive and love withdrawal methods seems to produce children who act on the bases of the fear of punishment or the avoidance of anxiety. These findings indicate that perceptually based mechanisms – assisted by instruction from the care giver – are most reliable in the establishment of conscience. If this is so, and the evidence for it is good, then there is scientifically based evidence for a perceptual source for the reliable identification of a moral end.

On the other hand, we can argue that we have a solid theory of moral ends, for example, one of a mixed evolutionary and cultural sort, involving reproduction and survival as evolutionarily based ultimate values and such things as literature, art, and science as culturally based values. Helping a person in need can then be identified as valuable from the theoretical perspective of what constitutes the morally valuable. So, empathic motivation to achieve that end is a morally justified motivation. It is a reliable motivational mechanism, since it usually leads to actions to help those in need, and it is a morally justified motivational mechanism because the goal in question is morally valuable. In this case, the reliable mechanisms that lead to the identification of the theoretically identified basic values of survival and reproduction are a complex set of capacities and skills that enable successful empirical testing, predictive power, explanatory power, and coherence with other well-supported scientific theories.

To illustrate the integrationist's claim that we can identify moral ends and thereby reliable mechanisms for achieving moral ends, I have focused on one relatively empirical means by which we identify moral ends, empathetic distress, and, on the other end of the continuum, one theoretically based means, the use of evolutionary theory to identify reproduction and survival as moral values. Obviously, what I have done can serve only as a promissory note on the larger task of not only filling out the details of these two sorts of cases but also addressing each of the mechanisms for identifying and discerning moral ends that are part of the integrationist's model of moral agency. To the extent that my model of moral agency is well supported scientifically, I have already begun that task, but needless to say, it is only a beginning.

Identifying reliable mechanisms

Besides identifying moral ends, we need to identify reliable mechanisms for achieving these ends. What do we mean by reliable mechanisms, and have we really identified any of them yet? Reliability, of course, has to do with how well a mechanism works in doing what it is designed to do. Since no mechanism is suited to work on all occasions and in every circumstance, reliability must be defined in terms of the quality of performance in circumstances for which the mechanism was designed to work, what we might call its normal working conditions. These conditions, depending on the purpose of the mechanism and its design, might cover a wide or narrow range of circumstances and occasions. Reliability is most easily measured in terms of the proportion of successful attempts to actual runs, and that measure can be gauged either absolutely or in comparison with mechanisms of a different sort designed for the same purpose. Or it can be assessed in terms of strategic successes, that is, in terms of how well it performs in contexts that are especially significant for the agent. Although the assessment of actual performances is the common source for judgments about reliability, assessments of possible performances, given knowledge of the mechanism and the possible circumstances and occasions of performance, are sometimes necessary or helpful in judging reliability.

What about the "item" that is being judged for reliability? I have referred to the reliability of mechanisms and sometimes of processes. When we consider some of the items whose reliability I have said is high on the basis of current scientific psychological knowledge – for instance, learning by practice, modeling, and instruction – it is clear that I am referring to complex mechanisms and multiple processes that are functioning in a coordinated fashion to achieve a goal. Undoubtedly, learning by modeling, for example, involves many different mechanisms and processes that themselves have component parts; indeed, it is probably accomplished in several different ways by different sorts of complex mechanisms and processes. In addition, it should be clear that whatever mechanism is being assessed, we are concerned with the reliability of a kind of mechanism, not the reliability of an individual mechanism in a particular instance of its operation.

Given these rough characterizations of what the integrationist has in mind by reliable mechanisms, I now return to the question of whether I have actually specified any reliable mechanisms for achieving true moral beliefs, proper moral motivations, and successful moral action. Even someone sympathetic to the integrationist project, not to mention antagonists and separatists, might be wondering whether the mechanisms for forming moral beliefs and motivations and performing moral actions, although reliable, are just as reliable in achieving amoral or even immoral ends.

The integrationist can probably make the best case for reliability with regard to mechanisms of moral internalization. We have seen that inductive techniques are superior to those of power assertion and love withdrawal in the formation of conscience. Induction, in some cases at least, involves roughly the perceptual and empathic capacities of the child and the cognitive and behavioral skills of the care giver in a setting in which the nature and consequences of an action are perceptually ascertainable. Although there are many details to be worked out concerning the nature and function of the factors involved in this sort of successful learning, it does seem that the integrationist has identified an effective sort of direct mechanism for moral learning, for the acquisition of true moral beliefs, proper motivations, and successful moral action.

However, our skeptical question seems to have some prima facie force with respect to the mechanisms involved in learning by practice, modeling, and instruction and to those involved in the self-system. We can easily imagine a person acquiring false moral beliefs through practice, learning, or instruction. For instance, Joe learns at Skinhead camp by means of one or more of these mechanisms that he ought to beat up blacks, Jews, and homosexuals whenever he has the opportunity. Since he has acquired his belief by means of a reliable mechanism, does Joe have a justified, though false, moral belief?[8] Perhaps this case and similar ones can be assimilated to misperception. Joe's false belief is just a very egregious one, like that due to a mirage. From a naturalistic perspective, our capacities for vision, hearing, touch, and so forth seem to be designed for the production of approximately true beliefs. Since they are not infallible mechanisms, they do indeed fail on occasion, especially in circumstances in which they are not designed to operate well or when they are themselves, because of some intrinsic factor(s), not operating well. But we are not tempted to think of them as reliable mechanisms for indifferently producing either true or false beliefs. Similarly, we can imagine a person acquiring through practice, modeling, or instruction false moral beliefs. So, on this interpretation, these ways of learning are generally reliable in the production of more or less approximately true moral beliefs; but, like all generally reliable mechanisms, they sometimes fail.

Although the capacity for empathic distress is similar to our perceptual capacities, and thus might be modeled on the latter with regard to issues of reliability, I do not think that we can fit learning by practice, modeling, and instruction into the perceptual model. The analogy between the reliability of

209

perception, on the one hand, and the reliability of learning by practice, modeling, and instruction, on the other hand, is strained, especially in the case of instruction. The reliability of practice, modeling, and instruction seems generally to be lower than that of perception. Although each involves the use of perceptual capacities, they involve more complex mechanisms, and the complexity itself gives rise to a greater possibility of error. But, in addition, the complexity is indicative of multiple component processes, each with its own specific ends. That these complex mechanisms are dedicated to the achievement, for instance, of successful *moral* action seems unlikely because of the coordination required for such an end. It seems much more likely that these mechanisms, if reliable, are so in the achievement of whatever is being taught. Thus they are best described as morally neutral reliable mechanisms, about as able to achieve immoral ends as moral ones.

I think that the integrationist has to concede that the level of specificity at which these mechanisms have been characterized thus far does not allow her to characterize them as reliable with respect to only moral ends. However, I do not think that this concession spells the end of the integrationist project. First of all, it should be recalled that we have adopted a well-supported theory of agency as such and applied it to moral agency on the supposition that generally reliable mechanisms should also be reliable in the moral realm. I think that this assumption remains a plausible one despite the difficulty that its lack of specificity raises for moral agency. One way that difficulty can be attenuated is by further attempts to specify more carefully the particular mechanisms and processes that go into successful moral learning. Second, it might be argued that although, with respect to all sorts of beliefs, practice, modeling, and instruction often lead to false beliefs, the reason for this is not the unreliability of the mechanisms but the content of what is being learned. This suggests, third, that a dynamic and contextual account of reliability is needed. Roughly, we might consider base-level mechanisms more reliable in the achievement of their ends in contexts similar to those in which they were formed either in evolutionary, developmental, or (noncognitive) learning history but less reliable in contexts that depart from these "normal situations." In these latter situations, higher-level mechanisms are required in the achievement of the same ends, those that promote reproduction and survival, for example, as well as for the achievement of moral ends that are nonevolutionary in character. But the ability of these higher-level mechanisms to operate in many more circumstances than the base-level mechanisms, as well as their adaptability to new circumstances, makes them less reliable. Moreover, higher-level mechanisms – for instance, learning by modeling and instruction – are more pervasively mediated by moral instructors and, thus, are subject to the vagaries of moral background beliefs, motivations, and practices. Finally, it is plausible to suppose that in these more complex circumstances true moral beliefs, proper moral motivation, and successful moral action are more difficult to achieve.

Lastly, what about the mechanisms of the self-system, the ones associated with the self-referential and self-reflective levels of our model of moral agency? One of the reasons these higher levels of moral agency are developed is precisely that they often better ensure the achievement of moral ends than do the base and behavioral levels operating alone.[9] The reliability of the self-reflective level depends to a large degree on the adequacy of the moral standards it employs, whereas the reliability of the self-referential system is very much dependent on the adequacy of the person's conception of herself as a moral agent. But, in both cases, given that capacities for employing standards and self-conceptions are reliable mechanisms for the achievement of goals generally, it seems plausible to argue that in their application to moral ends they are also effective. Nevertheless, it is clear that much more needs to be done in understanding the formation and use of these mechanisms. That increased understanding will, hopefully, bring a better estimate of their reliability both generally and contextually.

I conclude, then, that the integrationist is not completely unable to identify reliable mechanisms of moral agency. Although some of the mechanisms of moral agency, especially those of learning by practice, modeling, and instruction, are, as specified, general mechanisms of agency, reliable for amoral and even immoral tasks, others, like empathic response, coupled with inductive techniques, are identifiable as reliable moral mechanisms. The moral mechanisms of the self-referential and self-reflective levels share the reliability of their "amoral" counterparts while varying in reliability with the adequacy of the moral standards and conceptions of the moral self that they endorse.

8.8 A SCIENTIFIC NATURALISTIC ACCOUNT OF THE JUSTIFICATION OF MORAL AGENCY AND THE NATURALISTIC FALLACY

We have seen that in order for the integrationist to give an adequate account of moral agency, she has to answer four central questions about moral agency, those of relevance, acquisition, action, and adequacy. Although providing satisfactory answers to the first three of these questions is no mean accomplishment – and, indeed, I do not pretend to have provided anything more than some plausible hypotheses about answers to these questions – it is clear that the question of adequacy presents the greatest challenge to an integrationist. For it is the thesis that the descriptive and explanatory, on the one hand, and the prescriptive, evaluative, or normative, on the other hand, are linked that is the most controversial one. It is this thesis that runs afoul of the naturalistic fallacy and, thus, appears doomed to failure.

Roughly, we can specify two *forms* of the naturalistic fallacy: a *definitional form* that involves the illegitimate identification of values with facts, brought to philosophical awareness and scrutiny especially by G. E. Moore, and a *derivational form* that involves the illegitimate derivation of values from facts, pungently ensconced in philosophical consciousness by David Hume. It is helpful to distinguish not only definitional and derivational forms of the naturalistic fallacy but also two *versions* of the naturalistic fallacy, an epistemological version and an ontological version, corresponding to issues of justification and the nature of values, respectively. Each version has a definitional and a derivational form. The epistemological version of the fallacy involves confusing, and so illegitimately connecting, the sources of the acquisition of beliefs, on the one hand, with good reasons for claiming that one's belief is justified, on the other. In its definitional form, the epistemological version of the fallacy concerns confusing more immediate or direct *causes* of beliefs about values with more direct or immediate (basic) justificatory *reasons* for accepting a moral belief; in its derivational form, it takes mediate *causal sources* for moral beliefs for mediate (nonbasic) *justifiers* of moral claims. In each case, causal explanations are confused with or substituted for justifying reasons. On the other hand, the ontological version of the fallacy, as we shall see in more detail in the next chapter, involves, in its definitional form, the confusion and illegitimate identification of natural facts, properties, or characteristics with moral facts, properties, or characteristics; and, in its derivational form, it involves the claim that natural causes bring about moral facts.

Since I have been concerned with issues of justification in this chapter, let's focus on the epistemological version of the fallacy for the moment. Does our proposal commit the epistemological version of the naturalistic fallacy in either or both of its forms? I do not think so. The integrationist grants that there is a legitimate distinction to be made between explanation and justification and that not all causes are reasons. But although the integrationist concedes that not every acquisition mechanism is reliable, she argues that some are, and that there is no fallacy in appealing to these sorts of mechanisms in the justification of her moral beliefs. In some cases, the justificatory mechanisms may be more or less direct and noninferential, for instance, as in the case of empathically based responses; and, in others, they may be more or less indirect and inferential, as in the case of evolutionary theory and the theory of natural selection. Since the sciences, in particular biology and psychology, identify and characterize the workings of these reliable mechanisms – or at least clarify and give increased understanding to mechanisms that are identifiable and characterizable in some fashion by ordinary, nonscientific means – the integrationist's appeal to scientific theories and understandings is not fallacious; indeed, it explicates and explains the justificatory process. Thus the integrationist claims that even if there is an epistemological version of the naturalistic fallacy, she has not fallen afoul of it.

But what about the ontological version of that fallacy? In carrying out my scientific naturalistic project of the justification of moral beliefs, motives, and actions, I have made some realist assumptions about the nature of justification. I have characterized what traditional epistemology has called good reasons that provide justification for beliefs as reliable mechanisms for attaining the right sorts of ends – in the case of moral beliefs, approximately true moral beliefs. At one point in this chapter, I have taken a *true* belief to mean something like a belief that corresponds to what is the case. Moreover, I have discussed reliable moral cognitive mechanisms as those that put us into epistemic contact with moral properties of things and events, and I have described reliable moral motivation mechanisms as those that enable us to instantiate morally right and good actions. Thus, I have made ontological assumptions in the development of an integrationist hypothesis about the answer to the epistemological question of adequacy. I have maintained that I have presented a plausible hypothesis about epistemic justification while responding sufficiently to separatist and antagonist critiques that I have made some fundamental error by not attending to the epistemological gap between explanation and justification. Nevertheless, I would be indulging in wishful thinking to conclude that I have completely satisfied these critics. Indeed, critics might well claim that by linking explanation and justification, I have only dug myself into a deeper hole. In other words, critics might maintain that the naturalistic epistemological rope that I am using to try to bridge the fact/value chasm is made up of fatally flawed ontological fibers. Integrationists, proceed at your own imminent peril! Well, as they say, it's time to take the plunge and move on to an explicit consideration of ontological issues.

NOTES

1. There are a number of problems with this definition, but we do not need to address them directly in order to fill out the connections I want to make between justifications of moral beliefs and justifications of beliefs in general.
2. This approach contrasts with that of noncognitivism. Noncognitivists deny that morality is a matter of justified true beliefs about morality. That is not to say that they don't think that we can have beliefs about morality. Rather, it's that instead of moral beliefs being at the core of the moral agent's moral capacities, there are other agent-side factors, such as attitudes, moral preferences, or emotions, that are crucial in moral assessment. In the noncognitivists' view, then, moral assessment is not concerned with whether a moral agent has justified moral beliefs or justified true moral beliefs. Moral assessment is concerned with the rightness or correctness of the pertinent agent-side factor(s). For instance, one form of ethical noncognitivism is called *prescriptivism* because it holds that ethical propositions are commands rather than assertions. Commands, of course, are neither true nor false because they are incapable of truth or falsity. One classical method of assessment for a prescriptivist is to ask herself whether her proposed course of moral action is of the sort that anyone in the same situation ought to carry out. If so, then the proposed action is morally right.
3. The coherence theory of truth asserts that a belief or proposition is true if it coheres with other beliefs or propositions. The general idea behind the pragmatic theory of truth is that a belief is true if it works. All versions incorporate the idea that a belief works if it is successful in doing what it is supposed to do or achieving what it is designed to achieve.
4. The views of Peter Singer, Michael Ruse, and Robert Richards illustrate important differing stances on this issue. All three are sympathetic to sociobiologically based accounts of moral agency but have significantly different positions with respect to the potential of science in general, and sociobiology in particular, for answering the question of justification. Moral philosopher Peter Singer (1981), though, who unlike many other philosophers is appreciative of the insights that sociobiology offers moral philosophy about the biological causes of moral agency, argues for a position that is close to the traditional separatist position. He contends that Wilson's use of sociobiology cannot provide the justificatory reasons that moral justification demands. Historian and philosopher of biology Michael Ruse (1986) maintains that although sociobiology can provide no ultimate justifications for morality, there are in fact none to be had from any source. In Ruse's view, all that we can expect in the end are causal explanations of moral agency, and there sociobiology makes a big contribution. Finally, Robert Richards (1987), also a historian and philosopher of biology, argues that we can expect not only causal accounts but also justifications from a Darwinian evolutionary theory that has been enriched by the findings of sociobiology. Their analyses are helpful for understanding the important possible positions on the question of the role that scientific findings can play in ethical justifications, with

214

respect to not only the specific contributions of sociobiology but also those of biology generally as well as psychology. I have examined their views from an integrationist's perspective elsewhere (Rottschaefer and Martinsen 1984; Rottschaefer 1991a,b).

5. I have described the function of each of these levels in relatively separated fashion. However, a more realistic account would have several or all of the levels functioning together. An account of successful functioning embracing several or all levels would require adequate descriptions and explanations of the interaction of the levels cognitively and motivationally.

6. The model for empathic distress that Hoffman develops is called a *bystander* model because the situation envisioned is one in which the person in distress is present to the empathizer. Of course, the needs of persons are not always apparent to the potential empathizer, nor are victims themselves always present. However, empathic affect can be aroused in other than bystander situations because of humans' cognitive capacities. Hoffman suggests three ways in which this can occur. We are capable of representing events that can evoke empathy even in the absence of empathy-invoking stimuli. We can also imagine harmful stimuli affecting others. In addition, the semantic meanings of events can themselves become conditioned stimuli for autonomic arousal of empathic affect. In these ways, the cognitive representations of the behavioral and reflective levels can become attached to the empathic motivations of the base level.

7. The self-system need not always exercise its influence in an explicitly reflective manner, let alone in some deductively logical fashion. Nor is the self-system closed to influence. In Bandura's view, agency is highly situational, not only in the sense that what one does often depends on situational factors just as much as on internal factors, but also in the sense that it ought to do so.

8. We must, of course, distinguish this sort of case from one in which Joe has a justified belief that it is morally wrong to beat up blacks, Jews, and homosexuals but acts in a manner contrary to his belief.

9. That is not to say that the self-referential and self-reflective levels of operation are always to be preferred, at least, if we are sensitive to Bandura's cautions about the misuse of moral principles and an agent's capacities for self-exoneration. See Chapter 6.

215

9

Moral ontology

What is the ontological status of moral values?

9.1   FROM EPISTEMOLOGY TO METAPHYSICS

In the previous chapter, I was pursuing epistemological issues, in particular, an integrationist model for the justification of moral beliefs. One of my major claims was that justification should be understood in terms of reliable mechanisms that enable the acquisition of approximately true moral beliefs. We can take it for granted that if we hang around with epistemology for very long, her companion metaphysics will show up. In fact, issues of truth are ones that often bring metaphysics around. So it will behoove us, for a number of reasons, to turn now to the metaphysical issues that have been lurking around our epistemological discussions.

Although none of the classical views on truth we have examined thus far are necessarily antithetical to a realist view, the correspondence view is the one that seems to fit best with a realist position, that is, roughly, the view that true beliefs correspond with a reality that is independent of the ways that we might conceive of it in our thinking or theorizing and independent of the ways in which we might desire or wish it to be. Although one might rightly suppose that integrationists have a certain affinity for both scientific realism and moral realism, it is easy to see that that combination is not the only one available to an integrationist. Nevertheless, I adopt a position that embraces both scientific and moral realism. The issues concerning realism, whether in science or morality, are long-standing, complex, and far from resolved. My discussion is merely an introductory one whose purpose is to spell out one version of a realist integrationist hypothesis and give some indication of its plausibility. Consequently, for the most part, I take for granted the plausibility of arguments for scientific realism.

First, in Section 9.2, I lay out the territory of major realist and antirealist positions in morality. In Section, 9.3, I examine what moral reality looks like from the perspective of integrationist moral realism. Having laid out the integrationist moral realist's position that moral phenomena are natural phenomena constituted by natural factors, I proceed to defend the realist position in Section 9.4. First, I meet the challenge that integrationist moral realism is fundamentally flawed because it falls prey to an ontological version of the naturalistic fallacy. Having shown that integrationist realism is not ontologically impossible, I proceed to make the empirical case for a strong form of moral realism: emergent realism. In Section 9.5, I pull together the metaphysical discussion of this chapter about the existence and nature of moral facts and connect it with the naturalistic account of moral agency that I have developed thus far.

9.2 MORAL REALISM AND ANTIREALISM

As I have indicated, realism has something to do with the existence of realities that are considered to be independent of our beliefs or desires. Roughly, whether something is the case or not does not depend on whether we believe it to be the case or want it to be the case. Of course, this does not mean that there cannot be any such things as beliefs. My having an occurrent belief depends on my actually undergoing the process of believing. Does realism require that the existence of my belief is independent of whether I believe that I have a belief? This is a hard question and, fortunately, we will not have to try to answer it. However this question is answered, realists do not have to deny the existence of minds and of mind-dependent entities. Realism also means that reality isn't a certain way just because I want it to be that way. But clearly, at least at the folk psychological level, what we want and believe influences how we act, and sometimes we can make the objects of our desires and wants reality. Moreover, the coming to be of certain parts of reality depends, to some extent at least, on our beliefs and desires. Its continued existence is less, if in any way, so dependent. For example, while we are building a house, its coming to be depends on our beliefs, desires, and intentions about it. So too, if there is a divine creator of the world and the world continues in existence because of her continuing knowledge and will that it so perdure, then in a radical sense, created reality is mind dependent. Happily, once again, we avoid these sorts of complications. So, I do not try to develop a general definition of realism that provides necessary and sufficient conditions for realism and that covers all matters of inquiry, including the kinds of questions I just touched on. I adopt a more modest approach, trying to formulate a description of realism that is satisfactory for our purposes.

217

As a start, to be a moral realist, one has to claim that (1) there are moral facts,[1] (2) there can be some true moral beliefs, (3) there are some (approximately) true moral beliefs,[2] and (4) the truth of these beliefs is not constituted by the kind of warrant they receive, but rather is constituted by their correspondence with moral facts.[3] Thus, moral realism excludes moral nihilism, moral skepticism, and moral constructivism.[4] These latter positions are nonrealist or antirealist.

But these four theses are not sufficient to distinguish the sort of integrationist realism I will be exploring and defending from other realist positions. In addition, integrationist realism holds that (5) moral facts are natural facts rather than nonnatural or supernatural facts; (6) these moral facts are relational; and (7) these natural moral facts are constituted by and brought about by natural phenomena, that is, phenomena explained by the sciences.[5]

Now that I have presented the integrationist's theses about the existence and nature of moral reality, we can begin to examine those theses in detail and see what kind of support they have. I shall begin with Theses 5, 6, and 7, which concern the nature of moral reality. Once we have a better understanding of what the integrationist thinks moral phenomena are and why she does so, we can examine her views about the existence of moral facts and the truth of moral claims, views embodied in Theses 1 through 4.

9.3 INTEGRATIONIST MORAL REALISM

Natural moral facts

To what sorts of moral facts does a realist integrationist account of moral agency commit the integrationist? On the one hand, we have moral agents who, according to my model of moral agency, possess a complex four-level capacity, each level of which has its own structures and processes. On the other hand, we have talked about certain sorts of paradigmatically moral action, like helping someone in need, and various sorts of values, like food, clothing, shelter, companionship, and development of capacities. What more is there to say? Aren't those the moral facts to which the integrationist theory of moral agency refers, and if true, successfully refers? Yes, but this is only the beginning of the task of laying out the realist integrationist view on moral facts. There are some further important questions that need to be answered. Specifically, we need to answer the following three questions: (1) What is the nature of the moral facts referred to by the integrationist theory of moral agency? (2) In what way are these moral facts relational? (3) In what sense are these allegedly moral facts moral?

Moral ontology

All the facts referred to earlier are natural facts in the sense that they are the objects either of inquiries in the natural or social sciences or of reflective common sense. This is an important thing to notice, since it allows us to distinguish explicitly between three different sorts of moral realism. The naturalist maintains that moral facts are natural facts in the sense that we just discussed. The other two sorts of moral realism are also distinguished in terms of the kinds of facts that are the alleged referents of their respective moral theories. They are supernatural facts and nonnatural facts. The former are the kinds of facts referred to by theological ethical theories. For instance, one might have in mind a generic sort of Christian moral theory that refers to the will of a creator God in the determination of what is morally right and wrong. That sort of ethical theory involves claims about supernatural facts.[6] We have also met the nonnaturalists in our earlier discussions. G. E. Moore is the champion of nonnaturalist facts or realities. He argued that ethical values are nonnatural in character and are grasped by a special faculty of ethical intuition. In Moore's view, the good is exactly such a reality. We are now ready to examine Thesis 6, the claim that these natural moral facts are relational.

Relational moral facts

There are several different views on the nature of natural moral facts. The differences appear in terms of the sorts of moral facts with which moral realists contend true moral claims correspond. We can think of moral facts as being objective or not. Objective moral facts would not include anything about moral subjects, for instance, their individual or commonly shared beliefs or desires. Classically, Plato's form of the good has been considered the paradigm case of an objective moral value and so a moral fact. *Objectivists* contend that all moral facts are objective in this sense. Nonobjectivists are either (1) *subjectivists,* who maintain that moral facts are facts about the individual moral agents only, or (2) *intersubjectivists,* who claim that moral facts are facts about some group of moral agents — perhaps all moral agents. Clearly, though, we might conceive of a fourth position that holds that moral facts are about objective, subjective, and intersubjective facts. I call this the *relationalist* position. For reasons that will soon become apparent, the integrationist adopts a relationalist position.

One simple way to understand the relational view of moral facts is to consider the relational character of the good. The integrationist contends that x can only be good for something; x cannot just be good. That does not mean that there are no intrinsic values; it just means that if something is intrinsically valuable, it is so because it has a special connection with a

219

certain sort of agent. That connection may be due to the agent's evolution-ary, developmental, or learning history, depending on how strong a notion of intrinsic values the integrationist has in mind. Since intrinsic connection has often been associated with essences, we might link intrinsic with an evolutionary connection, even though, given what we know about evolu-tionary theory, it is no longer possible to think, as Plato and Aristotle did, that there are essences and thus that there are sharp dividing lines between different kinds of entities. For the integrationist, explanations of moral agency require a relational account of values, since they all involve the notion of beneficial consequences for an agent or agents.

The agency that we are concerned with is moral agency. We have seen that there are both substantive and functional requirements for what counts as moral agency. The substantive requirements have to do with the kinds of actions that are within the moral realm. I have used as our paradigmatic case of substantive morality helping someone in need. Unless one is a moral nihilist, one will be willing to concede that although the boundaries of the substantively moral are fuzzy, some actions are clearly within the moral realm. We can use these actions to understand the objective side of moral facts, the kinds of end states, toward which moral actions are aimed and that successful moral action achieves. I have also claimed that there are func-tional requirements for moral agency. Here too the boundaries are fuzzy, although we have argued that actions based on the reflective level are clearly in the moral realm and that actions stemming from the behavioral or base levels alone may or may not be, depending on their causal history. For instance, the spontaneous substantively morally right deed of the virtuous person may count as functionally moral. Thus, given a substantively moral right action, we can discern for each level of moral agency a relational triad of (1) an agent acting at a given level of agency (base, behavioral, reflective, or self-referential); (2) the objective state of affairs toward which the action aims, the instantiation of which constitutes the moral good in the situation; and (3) the particular set of causal relations that connect agent and state of affairs. Granting that there are four sets of relational triads corresponding to the four levels of moral agency, let us say that the primary relational moral triad concerns (1) the moral agent acting in a morally correct fashion at the reflective level, (2) those objective states of affairs that constitute the moral good for the agent in particular situations, and (3) the causal relations that bind agent and objective state of affairs together.

The states of affairs constitutive of moral goods with respect to the agent make up the objective side of moral facts. Regarding evolutionary, develop-mental, and noncognitive learning history, these states of affairs are con-nected in intricate ways, and more or less tightly, to the values of survival and reproduction of the agent and others. But there is an intrinsic indeterminacy about the kinds of things and states of affairs that function in the achievement of these values. What constitutes human fitness is *intrinsically* indeterminate in at least two important ways. First, human fitness does not, for instance, determine the *particular* biological goods whose achievement enable it. Although we must have food and protection from cold in order to be fit, the particular foodstuffs and type of clothing are not specified merely biologically. They are also determined by culture. In addition, the scope of human fitness as a moral value does not seem to be determined biologically. It is not clear, as we have seen, that the evolutionarily based value of human fitness includes all humans. Second, human fitness is *extrinsically* limited as a value in several ways. The Darwinian naturalist need not claim that human fitness is the only value or necessarily the highest of a variety of naturalistically identifiable values. She is a value pluralist. She holds that in addition to the primarily biologically based value of human fitness there are other, primarily culturally based human values, for instance, science, mathematics, literature, and art. Furthermore, she can claim that there are other, nonhuman biological values, for instance, the continued existence and flourishing of nonhuman organisms.

But granting both the intrinsic indeterminacy of the value of human fitness and its extrinsic limitations by both primarily cultural human values and biological, nonhuman values, human survival and reproduction are necessary conditions for both cultural human values and culturally specified biological human values. As such, human survival and reproduction are not mere instrumental values, that is, ones that can be set aside in the achievement of some goal. So, in this narrowly conceived fashion, our claim about the objective value of human fitness and human survival and reproduction is not completely unwarranted. For their value is supported by the fact that they are necessary conditions for the achievement of what we deem to be intrinsically, or at least independently valuable and because they are often valued in themselves.

On the other hand, in the case of those moral goods that are the result of our reflective and self-referential capacities, we have moved beyond the causal relationships involved in evolutionary, developmental, and noncognitive learning history to those formed and maintained through both individual and cognitive social learning history. Although seemingly dependent on the moral goods constituted through the former sorts of relationships, these moral goods both expand the scope of moral agency and are constitutive of states of affairs that move far beyond those that make up the earlier sorts of moral goods. This brief look at some of the finer structure of the triadic relationship that constitutes the moral realities postulated by the integrationist gives us a hint of the richness of that reality and opens up large areas for further study and reflection. It suffices, however, to make the point about the relational character of the moral values postulated by the integrationist. We are now ready to explore the last thesis on the realist list that concerns the nature of moral facts, Thesis 7, the claim that natural moral facts are constituted by and brought about by natural phenomena, the phenomena studied in the sciences.

The natural constitution and construction of moral facts

The integrationist claims that moral phenomena are natural, as opposed to supernatural or nonnatural, phenomena. Moreover, they are relational natural phenomena involving agents and the objects of their activities, as well as the complex causal interactions that bring about and constitute these activities. Thesis 7 of the integrationist's realist program makes this claim explicit, contending that the constituents and sources of moral phenomena, both of the moral agent and of moral values, the subjective and objective sides of the moral phenomena including their causal relationships, are natural phenomena. One way to explicate and assess this claim is to examine the theories that form the bases of the integrationist's account of moral agency. The integrationist's contention is that these theories explain both the nature and the source of moral agency from a completely naturalistic perspective. That is to say, they make use of standard explanatory patterns found in scientific disciplines, disciplines that are prototypical examples of those concerned with studying the nature and causes of natural phenomena.

We have discussed two fundamentally different, although not necessarily opposed, ways of understanding the nature of explanation. The first view of explanation, deriving from the logical empiricists, makes it a matter of deductive derivation of a conclusion about the explanandum from premises about the explanans and initial conditions. This is the so-called deductive–nomological model of explanation. A second view interprets explanation in a causal fashion. Explanations concern causal factors of various sorts and how they are operative in constituting or bringing about their effects. It is clear that these two views need not be incompatible, since it is often possible to formulate explanations involving causes and effects into deductive arguments. However, there are problems associated with the deductive–nomological model of explanation that make some alternative preferable (Kitcher and Salmon 1989). The causal account is such an alternative, even though we need not claim that all explanation involves reference to causes and effects. Moreover, it seems prima facie to be the sort of account that fits the theories I have invoked in accounting for moral agency.

Once we have adopted this model of explanation, it is quite natural to want some clarification of the notion of causality. I adopt a broad notion of causality with classical roots that allows for different sorts of causality. Thus, I do not, as is often the case in contemporary discussions of causality, confine the notion of causal explanations to explanations in terms of something like Aristotelian efficient causes. I take causal explanations in a broader sense to include explanations in terms of constitutive factors and in terms of either efficient or final causes.[7] The latter sorts of causes are often distinguished as mechanical and teleological. This characterization is not perfect since both mechanical and teleological can be interpreted too narrowly. As I indicated in our earlier discussions of natural selection in evolutionary theory and environmental selection in behavioral theory, mechanical and teleological explanations can be interpreted in a broad sense to refer to explanations in terms of a phenomenon's antecedents and consequences, respectively. That is the way I shall interpret them. Thus, mechanical explanations are not confined to mechanics, as it is worked out in Newtonian mechanics, nor are teleological explanations confined to explanations of conscious, purposive human actions. Indeed, using the consequences versus antecedents idea, we can include in the former explanations of conscious, purposive human actions goal-directed behaviors and functions. So, for instance, we sometimes respond to the question "What is that robin doing?" with the answer "It is hunting for worms" or "What is that thing?" with "It's the water pump." We contrast these sorts of explanations with explanations that answer questions like "Why did the pump break?" and "How did the bird's wing get injured?" These questions are answered with appeals to antecedents: "He put too much pressure on the handle" and "It flew into the window." In the consequentialist explanations of a behavior, the consequences of the behavior are explanatory of the capacity for that behavior. In consequentialist explanations of a function, the consequences of a certain item explain that item's existing, being the way it is, or doing what it is doing. On the other hand, explanations of behaviors in terms of antecedents appeal only to the operation of factors antecedent to the behavior and what they effect. The bird flew with considerable force into the window and at something of an angle, so that its left wing hit the window. Or he exerted so much force on the pump handle that it just snapped off.

It will help to put two types of consequentialist explanations into schematic form. First, the teleological explanation of a behavior: Let S = a subject, B = a behavior, and G = a goal. Then we can say: S does B for the sake of G (or for attaining the goal G) if and only if (i) B tends to bring about G and (ii) B occurs because (i.e., is brought about by the fact that) it tends to bring about G (Wright 1976, p. 39). Applying this to our robin, we can say that the robin is hunting for worms if and only if (i) hunting tends to bring about the discovery of worms and (ii) hunting occurs because it tends to bring about the discovery of worms. Functional explanations have a similar sort of schema: Let X stand for the item in question and Z for the consequence it brings about. Then, we can say: The function of X is Z if and only if (i) Z is a consequence (result) of X's being there (e.g., existing, being the way it is, doing what it does) and (ii) X is there because it does (results in) Z (Wright 1976, p. 81). So, the function of this metal thing with the handle is to pump water if and only if (i) pumping water is a consequence of this metal thing with the handle and (ii) this metal thing has a handle because it results in the pumping of water.

Granting that there is a plausible distinction between explanations in terms of consequences and antecedents, what use does the integrationist have for this distinction? Thesis 7 asserts that one major reason why moral phenomena are natural phenomena is that they are constituted of and constructed by natural phenomena. One way to show that they are so constituted and constructed is to demonstrate that moral phenomena are explained in the same way as other, nonmoral natural phenomena are explained. The integrationist's agenda is to show that consequentialist explanations of a particular form serve to explain moral agency as a complex capacity for agency that is itself constituted by four levels. The particular form of consequentialist explanations is that the consequences that serve as causal factors in the explanation of moral agency are moral values. So, we can call these sorts of explanations *good-consequence explanations*. The explanations are *etiological* in the sense that they account for how the phenomena in question came about. They are *consequentialist* because the factors doing the accounting include specific consequences of the phenomena in question, and the explanations are *good* consequence explanations because the particular consequences are moral values.

It is also important to notice the nature of the explanandum in question. The integrationist cannot be satisfied with explaining individual behaviors or types of behaviors, or individual functions or types of functions. Rather, she is interested in the explanation of capacities. Capacities are, of course, manifested through behaviors; but not all behaviors manifest capacities. Medieval philosophers had a saying: The inference from something's being to something's being able to be (or the inference from manifestation to potential) is a valid one. But the potencies or dispositions to act that we are concerned with in moral agency are more permanent, long-term ones than the occurrent, short-lived disposition that might result, say, from a brief bout with the flu, depression, or winning a jackpot. Nevertheless, the discovery of goal-directed behaviors plays an evidential role in the ascrip-

224

tion of the presence of capacities since behaviors are manifestations of capacities (not forgetting, of course, that they are also ways in which capacities are acquired and maintained). So, the integrationist attempts to establish Thesis 7 by showing how moral phenomena, in particular the capacities of moral agency, are explained in terms of good-consequence etiology.

We need to consider one more preliminary, and then we will be able to proceed with our task. In Chapter 2, I argued that natural selection is a kind of selection theory. It is helpful at this point to lay out in some detail what a selection theory is since it can provide a model – with modifications for some cases – for understanding the kinds of theories that integrationists claim provide explanations in terms of good-consequence etiologies. A selection theory contains the following elements (Darden and Cain 1989):

A. *Preconditions*
  1. A set of Y's exist and
  2. Y's vary as to whether they have the property P and
  3. Y's are in an environment E with critical factor F.
B. *Interaction*
  4. Y's in virtue of possessing or not possessing P interact differently with environment E and
  5. Critical factor F affects this interaction such that
C. *Effect*
  6. The possession of P causes the Y's with P to benefit and those without P to suffer.
  6'. This causal interaction may have the concomitant effect of sorting Z's.
D. *Longer-range effect*
  7. C, the effect, may be followed by D, the longer-range effect: increased reproduction of Y's with P or reproduction of something associated with Y's.
E. *Even longer-range effect*
  8. D may be followed by longer-range benefits.

When we examine the structure of selection theories, we can see that they have the structure of good-consequence etiologies. The possession of P's has consequences that are beneficial in various ways to their possessors – and perhaps to related others. Moreover, these beneficial consequences occur because Y's are in a certain selective environment with critical factor F, and the interaction of F's with Y's that have P is such that a benefit to the Y's occurs. Recall our example with the dark- and light-colored moths. The dark color of the moths served as a camouflage to protect them from predation from birds in an environment where the trees had been darkened by soot. The dark color was an adaptation that served a function in the given environment and thereby increased the differential survival and reproduction of the dark-colored moths with respect to the light-colored ones. Our example highlights one further point about the selection theory model. Two levels of explanation are implicit in that model. At the first level, the presence of the coloration is explained as an adaptation; it is a phenotypic character that serves the function of camouflage. The function explains the presence of the coloration by answering the question: Why are so many moths dark colored? But there is a further explanation for the adaptation; it promotes the fitness and, thereby, the differential survival and reproduction of the dark moths. The explanation answers the further question: What's the point of the camouflage? So, the first-level explanation provides a proximate good-consequence etiology, and the second-level explanation provides an "ultimate" good-consequence etiology.

Focusing now on moral agency, we can ask how each level of moral agency is acquired and maintained. In response to these questions, I have offered an explanatory theory that is selectionist in character. The base level of moral agency is explained on the basis of two important theories: the theory of natural selection from evolutionary biology and the theory of environmental selection from behavioral theory. Evolutionarily based moral capacities are the results of natural selection, and behavioral capacities are the consequences of environmental selection. Moreover, the cognitive, motivational, and behavioral capacities constitutive of the behavioral, reflective, and self-reflective levels of moral agency are arguably in part the result of what I call a *cognitive selection process*. All of these theories have the structure of a selection theory. I now look at the fine structure of the integrationist's account of the acquisition of moral agency to see if we can discern at each level of agency explanations in terms of a good-consequence etiology itself governed by selectional factors.[8] To put my point in a somewhat different way that brings out the role of the selectional factors, the integrationist claims that moral values play an explanatory role in natural selection, environmental selection, and cognitive selection. It will take two steps to lay out the explanatory role played by values in these theories. First, I shall maintain that these explanatory patterns reveal teleological processes, that is, processes governed by consequences. Secondly, I shall show that these consequences are moral values.

Using the selection theory, consider the capacity for empathic distress. We can make the following substitutions for the variables: Y's = humans or hominoids; P = the capacity for empathic distress; E = the natural and social environment of Y's; F = social arrangements and natural exigencies requiring cooperation; Z = genes and gene combinations responsible for the capacity of empathic distress. Finally, let the possible longer-range benefits accrue to groups having a large number of individuals with the capacity for empathic distress. In this case, nature has selected for organisms with a capacity for empathic distress. There is selection in the sense that only some of the possibilities have been actualized. What is it that explains the actualization of some of these possibilities and not others? Clearly, it is the consequences of these possibilities – in our case, that of evolutionarily based moral capacities or, specifically, the capacity of empathic distress. The consequences produced by the P's in the past account in part for the presence of the P in future generations.

This explanation is consequentialist in form and involves objective selectional factors in the explanation of the acquisition of the evolutionarily based moral capacity of empathic distress. Moreover, the consequences turn out to be either relatively more beneficial or less beneficial immediately for the individual and for those who are the beneficiaries of the helping behavior that often results from empathic distress. In addition, given the evolutionary context, these immediate results themselves have beneficial consequences for direct and indirect descendants, and perhaps for the group or population of which the individual is a member. Because some consequences play a role in the processes of adaptation precisely insofar as they are beneficial or harmful, we can say that the adaptive process is governed by values. These values are ultimately those of survival and reproduction. Of course, as we have seen in our discussion of survival and reproduction as moral values, it is not necessary to understand survival and reproduction as the highest human values, let alone the only moral values. The integrationist is a values pluralist. In addition, it is not necessary to consider survival and reproduction as intrinsic values; they may just be necessary means to moral values. Lastly, note that I am not addressing here the question of functional moral adequacy; that is, I am not claiming that a base-level action of agency that achieves a goal of empathic distress, for instance, helping behavior toward someone in need, is a moral action. I argued earlier that such an action probably is not functionally moral. My aim at this point is to examine the ends toward which each of the levels of moral agency are directed, and to explicate how explanations that invoke the capacities of each level have the form of good-consequence explanations.

Now let us see how this is also the case in the environmental selection of behavioral capacities at the base level of moral agency. Using the schema for a selection theory, we make the same substitution for variables, except that P's are operantly conditionable capacities for helping behavior. Several things should be noted about this instantiation of the selection theory pattern. What has changed is what is selected for; the consequentialist pattern and the fact that the consequences selected for are beneficial in such a way that they can be connected with moral values remains. However, it is not necessary that environmental selection be coordinated with natural selection; their ends may be consistent or inconsistent. After all, values, even moral values, sometimes clash or are irreconcilable. Moreover, the environmental selection process is distinct from the natural selection process, since the latter presupposes some sort of developmental connection between genes (or gene complexes) and phenotypic characters, in our case evolutionarily based moral capacities, a connection that is not required by what we called earlier nonevolutionary capacities. Since operantly acquired behavioral capacities need not be evolutionarily based moral capacities, although like all capacities they must be built on evolutionary capacities, we need not think of the selection process as including the effects described in 6', 7, and 8. But even if they do have these effects, the process by which they occur is not due to connections between genes and their phenotypic manifestations. I conclude that the environmental selection explanations of behavioral capacities subject to reinforcement, capacities that are included in the base level of moral agency, are value-based explanations. The moral values in question are the results of those critical factors in the selecting environment, whether persons, situations, events, or conditions, that interact with an organism or a person. As consequences of the organism's behavior or the person's action, they are enhancing for the person and others. These consequences may also be beneficial for offspring and for the population in an indirect fashion if there are connections between the behavioral consequences and survival and reproduction. Moreover, the environmental selection explanation of base-level behavioral capacities does not, as we have seen, require the introduction of conscious, goal-directed behavior. Positive and negative reinforcers do not necessarily have to function as consciously held goals and purposes whose consequences are anticipated to be beneficial and harmful. Thus, the explanation of the origin, development, and survival of certain behavior patterns by means of environmental selection, just like the natural selection of populations, instantiates a teleological and value-laden explanatory pattern, although it also does not involve conscious, goal-directed activity.

We are now ready to consider capacities that are operative at the behavioral, reflective, and self-reflective levels of moral agency. To illustrate, I focus on behavioral level capacities. The integrationist offers an explanation of these capacities in terms of a cognitive selection model. Let Y's be humans; let P's be cognitively acquired capacities for having beliefs, desires, and intentions about others, and let the other variables stay the same; but drop D from the schema. The central notion for understanding this application of the selection model is that of a cognitive environment. Because of our cognitive capacities, we can represent to ourselves both the external environment and the positive and negative consequences of our behaviors in that environment, even in their absence. Thus we can remember past behaviors and their consequences in various environments. We can note the consequences of our behaviors while we are doing them, and we can anticipate the consequences of our behaviors without performing them. Moreover, our cognitive capacities enable us to set up an internal environment that more or less accurately represents the external environment. Depending on how good our representations of behaviors and selecting environment are, we can thereby determine more or less accurately, before performing them, which behaviors will be successful and which not. So, without suffering the actual costs, we can weed out unfruitful behaviors, as well as the thoughts and desires that prompt them. Thus, we have the cognitive environmental selection of anticipated behaviors. This explanatory pattern, it should be clear, is both teleological and value-laden. Indeed, we are now at the level of agency in which, barring reductionistic or eliminativist anxieties, it is most appropriate to speak of explanations in terms of purposes, plans, and projects, and in particular, of moral agency. Since the selection of the behavior is governed by its anticipated consequences, the process is teleological. Moreover, these consequences are themselves either beneficial or harmful to the agent and others; and, precisely because of that influence, what occurs, the selection of the capacities for certain sorts of beliefs, desires, and intentions, is value-laden. Thus, cognitive environmental selection explanations are also good-consequence explanations.

As with the environmental selection explanations, I should note several caveats about this sort of explanation in relation to natural selection applications of the model. First, as with the realization due to environmental selection, it is not necessary to bring in any effects beyond 6. The cognitive, motivational, and intentional capacities selected for may have consequences that are consistent with evolutionary ends, but they need not. Moreover, we expect that these capacities would achieve consequences that move beyond those directly connected with the means for achieving reproduction and survival. Like environmentally selected capacities, these capacities are nonevolutionary, and although all capacities must somehow be linked to evolutionarily based ones, these links are not due to the genetically based processes that result in phenotypic characteristics.

There is another important difference between (some of) the capacities that constitute the behavioral, reflective, and self-reflective levels of moral agency and the base-level capacities. This difference calls for a significant change in the selection model. The variants on which natural selection operates when we are dealing with evolutionarily based moral capacities and basic, operantly conditionable capacities are not environmentally coupled. What this means is that natural selection does not involve the organism changing its capacities in order to fit in better with the selecting environment. Nor does the environment act directly on the capacities of a given organism in its lifetime to make the organism better suited to it. But this sort of coupling can occur in the case of cognitive capacities. For example, we can change our ideas in the light of what we know about the environment so that they may better help us achieve the ends that we believe are attainable in our current environment. So too can the environment induce changes in the variations that it will subsequently select for. For instance, it may give us new ideas about how to proceed in a situation that is more than usually hostile. But these departures from a selectional model to what some have called an *instructional* model do not alter the teleological or value-laden character of explanations that use it (Darden and Cain 1989). The subject-induced changes are guided by certain beneficial or harmful consequences, and the environmentally originated changes can be used by the subject to attain her ends.

Thus, the explanations that form the bases for the integrationist account of moral agency involve essentially beneficial and harmful consequences of the actions that flow from the capacities that are constitutive of the four levels that make up moral agency. Moreover, these explanations concern objective properties of things, events, organisms, and other human agents. These properties, insofar as they are beneficial or harmful to the agent, others in the population, society or the entire human community, are what the integrationist often refers to as *objective values*. That is not to say, of course, that the moral agent is not objective. Both the agent as the subject of moral capacities and activities and the realities with which the moral agent interacts are objective in the sense that their existence and natures are independent of the agent or someone else believing that they are so or desiring that they be so.[9]

I have been considering the kind of moral realism that distinguishes integrationist moral realism from other moral realist positions. The integrationist position is captured, we recall, by the following realist theses: (5) moral facts are natural facts rather than nonnatural or supernatural facts; (6)

these moral facts are relational; and (7) these natural moral facts are constituted by and brought about by natural phenomena, that is, phenomena explained by good-consequence explanations. Granted, then, that we have some understanding of integrationist moral realism, we still have to address the central issue of whether the sorts of moral facts advocated by the integrationist actually exist. We have yet to see how the integrationist supports Theses 1 through 4. These theses are, of course, intimately connected. Given a correspondence account of truth, if the integrationist can establish Thesis 3, that her claims about moral facts are true, then, of course, Thesis 2, that claims about moral facts can be true, and Thesis 1, that moral facts exist, both follow immediately. The connection between Thesis 4 and the first three theses is subtle. Thesis 4 asserts that the truth of moral beliefs is not constituted by the kind of support they receive, but rather is constituted by their correspondence with moral facts. Another way to put Thesis 4 is connected with our earlier claim that moral facts are objective in the sense of being independent of one's beliefs and desires. Thesis 4 adds to that the notion that these facts are independent of one's *justified* beliefs and desires. How is that sort of independence established? One way that it can be established is to show that the moral facts referred to by the moral realist play a causal role in effecting moral and nonmoral phenomena. It is the integrationist claim that this is in fact the case.

So, assuming that we have an understanding of the integrationist's realist claims about the nature of moral phenomena, we are now ready to address the question of the plausibility of these claims. Do such moral phenomena exist in the strong sense of the term that the integrationist is invoking? In establishing an affirmative response to this question, the integrationist realist is faced with two major challenges to her position. The first challenges the very possibility of her position, contending that moral phenomena cannot be natural facts; and the second argues that even if moral phenomena could be natural facts, there is no empirical evidence that they are. I shall address these two challenges in turn.

9.4 AN INTEGRATIONIST CASE FOR MORAL REALISM

Questions about the very possibility of moral facts

Nonnaturalists contend that integrationists commit the naturalistic fallacy by claiming that moral facts are natural facts. Even though antirealists do not share with their nonnaturalist colleagues the latter's commitment to nonnatural facts, they also reject the integrationist claim. Their antirealism is based on three prominent historical reasons: (1) the conviction that the fact–value barrier is unbridgeable, (2) the contention that appeals to supernatural facts are implausible because of the multiple difficulties in establishing the existence of transcendent divine realities, and (3) the lack of evidence for that illusive faculty of moral intuition that allows access to the nonnatural moral good. From the antirealists' perspective, then, antirealism seems to be the only option left. So, even though nonnaturalists and antirealists disagree about a lot of important issues, they agree that the naturalistic realism of the integrationist is her downfall. So, we have returned again to that old nemesis, the naturalistic fallacy, now in its *ontological* version: It is a mistake to make values, right and wrong, or more generally moral realities, part of the natural realm.

The ontological version of the naturalistic fallacy

In Chapter 9 we saw that each version of the naturalistic fallacy, epistemological and ontological, can be divided into two forms: the definitional and the derivational. This is a helpful way to look at the alleged fallacy since it puts the fallacy into the context of reductionism, which, as we remember, was concerned in the first place with the relationships of scientific theories and the terms of these theories. If we are scientific realists, then the presence or absence of reductions has ontological implications. Thus, one way to understand the naturalistic fallacy metaphysically is to see it as a prohibition against illegitimate reductions. The definitional form of the ontological version of naturalistic fallacy leads to the charge that the realist integrationist is identifying moral facts with natural facts, and the derivational form leaves the integrationist charged with trying to explain moral facts by invoking natural causes.

Since I have argued in Chapter 7 for an emergent mentalism with respect to moral agency, one can correctly assume that the integrationist is committed to a distinction between moral and natural facts. Nevertheless, there is a clear sense in which the integrationist is also dedicated to the view that moral facts are natural facts. For she claims that they are neither nonnatural nor supernatural facts, and that there are no sorts of facts except natural ones. Alas, must the integrationist now change her story and plead guilty as charged? As a card-carrying naturalist, the integrationist is not opposed in principle to the identification of moral realities and natural facts. If the empirical and scientific evidence demands it, then the integrationist needs to work out the reductions demanded by the empirical evidence and try to show that the so-called naturalistic fallacy is no fallacy at all. But from the point of view of some of her opponents, the integrationist's concerns about empirical matters are completely misplaced. For the gap between fact and value is not a contingent one. It is a necessary one. Thus, so the challenge goes, it isn't just a matter of moral phenomena not *happening* to be natural facts, but rather one of impossibility. In terms of reduction, the claim is that the reduction is impossible.

As we have seen, a reduction requires the fulfillment of both the definability and derivational conditions. The former requires that the central terms of the reducing theory be linked by means of bridge principles or correspondence rules to those of the reducing theory; in the case of moral theory, bridge principles could define moral value in terms of happiness, for example. The derivational condition demands that the reduced theory be deducible from the reducing theory. Taking up the latter demand first, it is easy to see that unless the reducing theory already contains terms that refer to moral phenomena or premises that so refer are introduced, deriving moral claims from nonmoral claims is logically fallacious. For instance, it is fallacious to derive from the premise that humans seek happiness the conclusion that humans ought to seek happiness. With respect to the first requirement about definitions, G. E. Moore (1903) offered what is called the *open question argument* to show that attempting to define values in terms of facts is fallacious. The argument claims that we (or any other competent speaker in our language) cannot assert that the moral good is defined in terms of nonmoral natural facts, for instance, pleasure, happiness, human fulfillment, survival and reproduction, or any other natural characteristic or process, because one can always legitimately ask: What makes that (pleasure, happiness, etc.) good? No definition in terms of a natural phenomenon will ever be able to answer that question. It always remains open.

However, it may be argued that even if moral and nonmoral terms are not synonymous, they may be related to each other by meaning implication. Moral terms may imply nonmoral terms or vice versa. But the proponent of the is–ought gap argues that if such meaning implications hold, then it should be the case that competent speakers would recognize such implications. So, we have another version of the open question argument extended from definitions to meaning implications. If moral term M necessarily implies nonmoral terms N or vice versa, then the questions (1) Is M necessarily N? and (2) Is N necessarily M? should be closed. That is, a competent speaker would answer affirmatively in both cases. However, so the extended argument goes, competent speakers will always find them open. They will not with certainty be able to answer them in the affirmative. Putting together the claims about synonymy and meaning implication, we have what has been called a *semantic* thesis (Brink 1989, Ch. 6). The semantic thesis renders the definitional connection impossible, whether the definitional connection is construed as stating a synonymy or a meaning

233

implication. The semantic thesis asserts that there are no analytic statements that by themselves, or in connection with synthetic statements, can be used to derive a moral statement. In particular, it denies the possibility of what we called earlier in our discussion of reduction the linking of moral and non-moral terms by means of bridge principles or correspondence rules.

So far, we have a semantic thesis about how one arrives at the meaning or meaning implications of terms by consulting the usage of competent speakers and a particular claim about how competent speakers would define moral goodness. Suppose that we also contend that such usage is indicative of the kinds of properties that are constitutive of things. The idea is that we can use synonymy or meaning implication relations between terms as a test of whether properties are identical or connected intrinsically, that is, by the ontological analog of meaning implication relations. Call this the *semantic test for properties* (Brink, 1989). Thus, combining the semantic thesis with the semantic test of properties, we can conclude that moral properties can be neither identified with natural properties nor constituted by them.

Confronting the naturalistic fallacy: The possibility of moral facts

Both the semantic thesis and the semantic test of properties are problematic. Competent speakers of ordinary usage are not the only sources for definitions. Definitions that reform ordinary usage are possible and, if justified, legitimate; in particular, definitions that reform the ordinary usage of moral terms may be legitimate and even required. Indeed, naturalists and/or integrationists argue that such reforming definitions, particularly naturalistic ones, are to be expected. In any case, the issue of definition should not be settled on the basis of merely competent ordinary usage. But the semantic test for properties is even more problematic, since it commits one to an a priori account of the nature of moral and natural properties in the sense that the criteria for the identity of moral and natural properties are based on competent linguistic usage, where that is understood to be a matter of linguistic convention.[10] For surely, the realist and naturalistic integrationist argues, the nature of moral values and realities is an empirical matter and not merely a matter of linguistic convention. Since the semantic thesis makes the definitions of moral terms a priori, it is inconsistent with an empiricist and realist approach to property identification and description advocated by the integrationist.[11] If the ontological version of the naturalistic fallacy in its definitional form presupposes, as it seems to, that the issue of the relationships between moral and natural phenomena is purely a linguistic, and thus an a priori matter, it surely begs the question. It presupposes an answer that rules out ahead of time a realist integrationist response to the question.

Consider the following case from Newtonian physics. What is force? Force is the product of mass and acceleration. Is that what force really is, you might ask? Well, you can ask the question. The question is meaningful and in this sense still open. But how good a question is it? At this point in the development of physics, it is not the best of questions. We seem to have a pretty good theory of mechanical force, one that includes the preceding definition. That does not mean that the definition might not have to be changed. But, on the other hand, it doesn't mean that force, in fact, cannot be identified with the product of mass and acceleration or something else. The same thing seems to be the case with the moral good. One can grant the continuing legitimacy of the question of what makes that really morally good without questioning the legitimacy of some particular identification of the moral good with some natural phenomenon, even if the identification is a provisional one, subject to revision, given good reasons for revision. To rule out before empirical investigation a certain account of the relationships between force, mass, and acceleration is to beg the question. So too with regard to naturalistic accounts of the moral good.

Moreover, if we accept both the semantic thesis and the semantic test of properties, since these are general theses about the determination of the definitions of terms and the nature of properties, then they apply to disciplines at all levels. They apply not only to the relationships between terms – and the properties to which these terms refer – in ethical theory but also, for instance, in social theory. In addition, they apply to the terms – and properties to which they refer – in all the various theories in disciplines at neighboring levels that we might be interested in relating, for instance, (1) the social sciences and psychology, (2) psychology and the neurosciences, (3) the neurosciences and biology, (4) biology and chemistry, and (5) chemistry and physics, just to name a few possibilities. Indeed, it appears that the open question argument applies to all attempts to relate any disciplines. Consider, again, the classical example of reduction, that of phenomenological thermodynamics to statistical mechanics. One of the key bridge principles in that reduction is the identification of the temperature of a gas with the average kinetic energy of its molecular constituents. Would a competent speaker assign the same meaning to "temperature of a gas" and to "the average kinetic energy of the molecular constituents of the gas"? Unless she would, the terms would not be synonymous. Are there meaning implications such that one can be derived from the other? It is not at all apparent that there are such connections. But if the terms are neither synonymous nor related by meaning implication, then, when the open question argument is applied, one question will remain closed and the other will remain open. A competent speaker would claim that the question "Is the temperature of a gas the temperature of a gas?" is definitely a closed one. But, on the other hand, she would claim that the question "Is the temperature of a gas the average kinetic energy of the molecular constituents of the gas?" is definitely an open one. The meanings of the subject and predicate terms are not synonymous and, using the ordinarily accepted meanings of those terms, one does not imply the other. The situation seems to be the same for other would-be bridge principles, for instance, that genes are strings of DNA molecules. Consequently, an ontological gap opens between the properties to which these terms refer.

Thus, even if the premises of the open question argument, that is, the semantic thesis and the semantic test of properties, are accepted, it turns out that the open question argument does not raise a special problem for relating ethical theory to theories in the natural and social sciences. The is–ought gap is just one of a number of other gaps between theories in different disciplines. These gaps just happen to be "is–is" gaps, but do not seem to be of a different character than that of the is–ought gap. Consequently, the antirealist in ethics, whether cognitivist or noncognitivist, is faced with the options of becoming an across-the-board antirealist or going back to the drawing board to reconsider the open question argument. So too is the nonnaturalist faced with the choice of reconsidering the open question argument or denying the possibility of unifying scientific disciplines, since each is closed off from any other by an unbridgeable gap.

Of course, the naturalist need not accept the premises on which the antinaturalists construct their argument. The semantic thesis is problematic in the sense that it restricts the determination of issues of synonymy and meaning implication to the usage of competent ordinary speakers. But the semantic test of properties is even more problematic because it makes the determination of ontological issues rely on relatively a priori methods, thus begging the question between the integrationist realist and her opponents. Consequently, the naturalist might accept the semantic thesis, admit on that basis that there is an is–ought gap, but then argue that such a semantic gap in ordinary usage has no bearing on the ontological question of whether moral properties are identical to or constituted by natural properties. There is no intrinsic barrier to moral phenomena being natural facts. The naturalist, who is also a realist and an integrationist, contends that the question of the relationships between moral and nonmoral phenomena is a matter for empirical research and theorizing, just as are questions about the relationships between the phenomena investigated in other disciplines.

Moral ontology

From possibility to actuality: The existence of moral facts

What might these relationships between moral and nonmoral phenomena be? Moreover, what justification does the integrationist have for her claims about these relationships? Are moral facts identical with some sorts of nonmoral natural facts, psychological states, or social configurations, for instance? Are they constituted by such states? I have addressed this question before in two different contexts: first, the nature of moral agency and, second, the nature of evolutionary values.

I argued in Chapter 7 that there is good reason to deny that moral agency is epiphenomenal and that an integrationist strategy need not be abandoned for an implausible separatism or an unsubstantiated eliminativism. I described this position as a strong emergentist position because it not only admits the presence of novel levels of reality that are multiply instantiated in lower levels, but also claims that these levels are causally efficacious. In Chapter 8, I addressed another part of this issue when I argued that some of the causal processes involved in agency, including the cognitive, motivational, and behavioral processes that make it up, reliably attain the goals of justified moral beliefs, proper motivation, and morally correct actions. I identified these causal processes as reliable mechanisms. As such, they are the justifiers of moral belief, motivations, and actions. Thus, besides the explanatory emergence that we discussed in terms of the emergent causal capacities of agency, there is a kind of justificatory emergence. The theses of explanatory and justificatory emergence approach the issues of relationships between natural moral facts and nonmoral facts from the side of the moral agent, the subjective side of the relational triad that constitutes moral reality for the integrationist. *Explanatory* and *justificatory emergence* refer to the moral agent and the types of capacities and causal processes that enable her to achieve moral ends. But, as we know, the integrationist also believes that there is a role for the object side, the states of affairs and things whose realization and attainment are constitutive of the moral good for a moral agent. This sort of role can be characterized in terms of ontological emergence. Of course, both subject and object belong in the integrationist's metaphysics, so in that sense the referents of both explanatory and justificatory emergence are ontological. Thus, we can separate out the role of the object side by speaking of it now in terms of *objective* emergence.

Moral facts as strongly emergent

What I want to establish is that the triadic natural relation that I have claimed is constitutive of moral phenomena – that of subject, object, and the

237

relationships connecting them – is a strongly emergent reality. So I now add to my original seven theses concerning moral realism one that asserts the strongly emergent character of moral facts: Thesis 8: Moral facts are strongly emergent in the sense that both the moral agent and the objective states of affairs that in their attainment are constitutive of the moral good for the moral agent are related in part by *causal* interactions. Consequently, we can attribute to both the subjective and objective sides of the relational realities that make up moral facts genuine causal efficacy. Theses 5, 6, and 7 concern the integrationist's view of the nature of moral phenomena, that is, that they are natural relational phenomena, constituted by natural phenomena open to empirical and scientific investigation. On the other hand, Theses 1 and 8 assert the existence of moral facts, the former what we might call their *bare existence* and the latter their *hardy* (that is, *causally active*) *existence*. Theses 2, 3, and 4 concern the truth of claims about moral facts. By establishing the truth of Theses 1 and 8, we can infer the truth of these latter theses. Moreover, since Thesis 1 follows from Thesis 8, I focus my attention on establishing Thesis 8. My strategy is as follows. First, I argue that the scientific theories on which my account of moral agency has been built should be taken realistically. Second, I contend that some of the empirical findings on moral development support the case for a hardy moral realism of the strong emergentist sort advocated in Thesis 8.

The scientific realist bases for the integrationist emergent moral realism

The integrationist contends that the sciences play a role in determining answers to the major questions of moral agency and has made use of evolutionary, developmental, behavioral, and social cognitive learning theories in the construction of an account of moral agency. An integrationist who is a realist in her metaphysical commitments interprets these theories in a realistic manner. That is not to say, of course, that either the integrationist or her colleague, the scientific realist, is committed to interpreting every scientific theory they use realistically and understanding all the entities and properties of entities referred to in these theories realistically. What the scientific realist and her integrationist moral realist colleague hold is that (1) scientific theories can be true and false, (2) some of them are approximately true, (3) those that are true are thereby referentially successful, and (4) the truth of these scientific theories is to be understood in a correspondence sense.

Although it is beyond the scope of my investigations to attempt to argue in detail for the preferability of scientific realism to that of antirealist understandings of science, it is nevertheless important for us to understand some of the reasons for the plausibility of the realist case and its preferability to its competitors. There are a number of different subtle antirealist positions, but for our purposes we should be aware of two (Leplin 1984). First, there is the internal realist, who identifies the truth of scientific claims with the conditions that make for warranted assertability and thus denies the scientific realist's fourth thesis. It follows from this that although the internalist is willing to talk about scientific entities and properties, they have no existence independent of the warranted claims by means of which scientists refer to them. The classical form of antirealism with respect to theoretical entities and properties is *instrumentalism.* Instrumentalists deny the scientific realist's first thesis and regard scientific theories as conceptual maps that help us get around in the empirical macroworld by enabling us to make good predictions about that world and to summarize a mass of data in compact form.

The arguments for scientific realism sometimes take the form of an inference to the best explanation. An inference to the best explanation is a nondeductive process of reasoning in which it is argued that a particular explanation of the phenomenon to be accounted for does the best job, relative to its viable competitors, in accounting for that phenomenon. Applying this to the issue of scientific realism, scientific realists argue that the best explanation of the success of the sciences is that scientific theories are gradually approximating an independent reality. This success can be measured in terms of the progress of theoretical scientific knowledge or the practical successes of the applications of scientific theories. A closely related argument for scientific realism contends that the best explanation of the successful features of scientific methodology (e.g., methods of measurement, detection procedures, principles of experimental design, choice of research problems, standards for the assessment of experimental evidence, principles governing theory choice, and the rules for the use of theoretical language) is a realistic one (Boyd 1984). Methodological practices, like the use of detection devices, are theory dependent. The electron microscope, for instance, depends on scientific theories for its construction, use, and interpretation. The reliability of the microscope is best explained by the relative reliability of the theories in terms of which it is constructed. In turn, the use of the electron microscope to test theories enables increases in theoretical knowledge. It is the reliability of such instruments in testing that helps to bring about this increase in theoretical knowledge. But, then, the increase in theoretical knowledge, in turn, enables the improvement on – and development of new – detection devices and other instruments. This sort of improvement and development of detection devices is best explained by the view that the theories on which the instruments are built are gradually approximating theory-independent realities. So, scientific realists conclude that the advance of scientific knowledge – through this method of using theoretically based observation to confirm new and different sorts of theoretical claims, which, in turn, enable the use of new sorts of observations based on newly acquired theoretical claims – is best explained by postulating the existence of scientific realities that guide both the theoretical testing and the "construction" of observations by means of theoretically based detection devices.

The integrationist argues that the case for moral realism is supported by the case for scientific realism precisely insofar as she has made use of well-substantiated theories to account for moral agency. Indeed, her position is bolstered when we reflect on the fact that the fourfold structure of moral agency is derived from well-supported theories in evolutionary biology, developmental and behavioristic psychology, and social cognitive theory. The integrationist argues that there are good reasons, given the earlier arguments and others for scientific realism, to interpret the theories on which her account of moral agency is based realistically. Therefore, there are good reasons to accept the reality of moral agency as we have characterized it.

Moreover, as we have seen, the theories of moral agency implicitly involve reference to realities independent of moral agency. That is to say, the objective side of moral reality must also be taken realistically. Or, to put it in another way, those things and states of affairs that are the objects of moral activity are themselves objectively morally good. The selectional character of several of the theories used in characterizing moral agency requires the reality of the objects of moral agency, the moral goods sought by moral agents. It is not possible to discuss natural selection, behavioral selection, or developmentally enabled adaptations without bringing in factors in the selecting environments that are partially responsible for the capacities and processes constitutive of the base level of moral agency. Moreover, insofar as the capacities and processes of the behavioral, reflective, and self-referential levels of moral agency can be accounted for by cognitive selectional models, a case can be made for the reality of the objective side of the relational triad at these levels of moral agency, since these theories presuppose that the cognitive variables that play a role in the selection of beliefs about ends and means – as well as the selection of motivations and intentions – are themselves dependent on the character of the situations in which the agent is to act. The *instructional model,* which seems to be required to account for some of the activities on the behavioral, reflective, and self-referential levels, also makes implicit reference to the guiding role of the realities of the situations in which the agent is present and must act effectively.

Thus, interpreting these theories realistically provides evidence for the hardy realism of Thesis 8 – and consequently for Thesis 1 – not only with respect to the subject side of the relational triad but also with respect to the object side. This is so because the theories on which the integrationist's account of moral agency rests are explanatory in nature. As a result, taken realistically, they attribute causal capacities to moral phenomena. However, the explanatory capacity of moral theories, specifically the explanatory capacity of moral theories concerning what we have called the objective side of the relational triad, has been challenged. I now turn to this challenge and use the opportunity of meeting it to develop another argument for emergent moral realism.

Moral development, moral internalization, and emergent moral realism

Discussions about the similarities and dissimilarities between ethics and science occurring in the heyday of logical positivism led to antirealist, noncognitivist views about moral claims. Logical positivists argued that scientific claims were testable and, because of that, had a bearing on reality; however, moral claims, they contended, were not testable. Consequently, moral claims could be neither true nor false and could not be about an independently existing moral reality. Moral expressions were, therefore, noncognitive in character. However, the logical positivists' account of how scientific claims are testable turned out to be incorrect. For they maintained that scientific claims are testable on their own. But it has become clear that scientific claims are not testable in that way. They require auxiliary hypotheses and statements of testing conditions in order to be tested. But then it turns out that if moral claims are similarly accoutered in the testing situation, they too can be tested.

Nevertheless, Gilbert Harman has argued that although moral claims can be tested, their testability and subsequent confirmation are not sufficient to establish moral realism (Harman 1977; 1985, pp. 27–48; 1986, pp. 57–68). The fact that an ethical principle has observable consequences does not, he contends, establish that there are moral facts. Thus Harman is contending that explanatory scientific hypotheses identify causal factors in the occurrence of the explanandum. Successful testing of such hypotheses is a basis for realism about these factors. So, if the moral realist is to make her case, she needs to demonstrate not merely that moral claims can be successfully tested, but also that some moral claims are explanatory hypotheses that, if confirmed, provide evidence that moral facts play a causal role in bringing about moral and nonmoral phenomena.[12]

But, Harman argues, claims like (M): *Helping someone who is injured and in need of help is morally right* do not have that sort of explanatory power. He maintains that subjective factors such as pro or con attitudes are relatively sufficient in bringing about moral action insofar as they do not require the objective rightness of actions in order to bring about their effects.[13] On this basis, he concludes that hypotheses about objective moral facts are not necessary in explanations of either other moral facts or nonmoral facts. Indeed, Harman's claim is even stronger; such hypotheses about objective moral facts are explanatorily irrelevant (Harman 1977).

Nicholas Sturgeon, among others, has challenged Harman's contention concerning the explanatory impotency of moral claims (Sturgeon 1985, pp. 49–78; 1986, pp. 69–78, 115–41; Railton 1986; Brink 1989). He has argued that commonsense moral explanations, if taken on face value, do attribute effects to the moral features of actions, persons, practices, and institutions. These explanations, he maintains, should be taken seriously and should not be set aside without good reason. Sturgeon contends that Harman has not presented any good reasons for not taking these explanations seriously and, a fortiori, for regarding them as explanatorily irrelevant. Of course, Sturgeon recognizes that commonsense explanatory practice does not provide definitive support for moral realism. So, he supports his argument with the contention that these explanations stand up in the face of thought experiments concerning counterfactual situations. For instance, suppose that Al hits his innocent baby brother with a stick, thereby harming him. Lisa sees Al doing so and forms the belief that what Al did was wrong. How do we explain her belief? Sturgeon argues that a necessary part of this explanation must be the fact that what Al did was wrong. He asks us to suppose that what Al did in hitting his innocent baby brother with a stick and harming him was not wrong. What, then, would Lisa have believed about what Al had done? Sturgeon contends that our intuitions tell us that she would not have believed that what Al had done was wrong.[14] So, he argues, there is some reason to believe that moral facts play a causal role in Lisa's belief.

Harman concedes that at least in some cases of ordinary moral reasoning the counterfactual test seems to work. But he does not concede that this is sufficient to establish Sturgeon's case for moral realism. He reminds us that the counterfactual argument would not be enough to convince a moral epiphenomenalist that moral properties are actually causally efficacious. They may exist, but they don't do anything. Moral epiphenomenalists would attribute all the causal efficacy to the nonmoral features of Al's action on which the moral features supervene. However, Sturgeon finds Harman's appeal to moral epiphenomenalism unsatisfactory because Harman's own naturalistic position itself depends on the failure of epiphenomenalism and because a full-fledged commitment to epiphenomenalism renders all scientific explanations, except perhaps those of particle physics, explanatorily inert.

I conclude that this discussion has reached a stalemate; neither the moral realist nor the antirealist has presented a clearly convincing case. Appeals to ordinary moral reasoning have run their course in the sense that examples favoring one side or the other are available and that all examples can be reinterpreted more or less well by the other side to fit its own hypothesis. In addition, thought experiments seem to lead to theoretical issues about the causality of supervenient properties postulated by social and scientific theories. However these issues are decided, it is not clear that the points raised for or against moral realism are particular to that thesis. In order to break this apparent stalemate, I return to the results of scientific psychological studies concerning moral development – in particular, those of moral internalization, which we made use of earlier – to see whether they throw light on the mechanisms of moral belief formation and thereby on the role, if any, of moral facts.[15] I contend that the current findings support a role for moral facts in moral belief formation that is closer to Sturgeon's realist account than to Harman's antirealist position. On that basis, I suggest that there is some empirical support for the moral realism favored by Sturgeon and others. That evidence suggests not only the existence of objective moral facts but also a causal role for these facts in the fixation of moral beliefs.

Moral internalization is one way that developmental psychologists often talk about what moralists have described as conscience and the development of conscience. It refers to a psychological state in which one feels or believes that one has an obligation to act in accord with moral norms, rather than for social approval or egoistic concerns. Developmental psychologists use the term *moral norms* in various ways; one acceptable and nonbiasing meaning is that a moral norm is a norm that enjoins one in a specific situation to act for the welfare of another. An agent manifests successful moral internalization when, in situations where there is a conflict of interests between the welfare of another and her own interests, she consistently acts to promote the welfare of another person, without regard for social approval or egoistic concerns.

We can interpret the background moral beliefs, motivations, and feelings assumed by both Harman and Sturgeon in their explanations of moral actions to be at least partially the result of moral internalization. What does the research on moral internalization tell us about the sources for these background beliefs, motivations, and feelings? I focus on the results of studying moral learning in children roughly 4 to 12 years old. Among the many factors that contribute to moral learning and internalization, the central ones are (1) the *child,* for instance, her age, temperament, and relative cognitive, affective, motivational, and behavioral capacities; (2) the *care giver(s),* their moral beliefs and motivations and their own moral learning histories; and (3) what we might call the *circumstances of moral learning.* These latter are often classified as (1) situations of *parental discipline,* (2) occasions for the child to *identify* with or *imitate* her care giver, and (3) occasions of *cognitive moral dissonance.*[16]

Situations of parental discipline are particularly apt for understanding the factors responsible for moral internalization. One reason for this is that situations of parental discipline bear important resemblances to those of moral conflict. In situations of parental discipline, the child's egoistic desires clash with moral standards that are "external," that is, standards that are embodied in the physical and verbal messages of the parents or care givers about how the child should or should not act. Situations of moral conflict provide one who has achieved moral internalization with the opportunity to engage her conscience in the resolution of the conflict. The moral problem is to resolve the conflict between one's egoistic – and presumably nonmoral or immoral – desires and the moral standards applicable in the situation in favor of the moral standards. Successful moral internalization manifests itself in the ability to act on the basis of these moral requirements without regard to external sanctions.

Psychologists who study moral development have identified three significantly different types of disciplinary methods by means of which parents facilitate moral internalization: (1) power techniques, (2) love withdrawal, and (3) inductive techniques.[17] Power techniques include such measures as the use of force, deprivation of privileges, threats, and commands. Love withdrawal techniques include expressions of parental anger and disapproval. In the use of inductive techniques, care givers point out to the child, either directly or indirectly, the effects of the child's behavior on others.[18] Studies since the late 1950s and early 1960s indicate that the most effective means of moral internalization are inductive techniques (Hoffman 1970; 1988). There is also some evidence that the occasional use of power assertion techniques by parents who usually employ inductive techniques plays a positive role in moral internalization by, for instance, letting the child know that the parent feels strongly about something or by controlling a child's defiant behavior (Zahn-Waxler et al. 1979).[19] In general, love withdrawal and power assertion techniques may get the child's attention, whereas inductive techniques indicate to the child, for instance, the harmful consequences of her actions. Inductive techniques help to engage the child's empathic capacities and can also lead to a feeling of guilt. The most successful moral internalization occurs by means of the repeated use of inductive techniques in varied circumstances of moral learning.

For example, consider Lisa's moral learning situation at a point when the moral internalization process is beginning under the auspices of her care giver(s). Even though at the beginning of the learning period Lisa is still a moral neophyte, her mind is not a moral blank slate, especially if the evidence concerning built-in empathic capacities is correct (Hoffman 1981, pp. 121–37). With regard to her care giver, we assume that no matter what internalization technique she uses, she considers the action in question wrong. Suppose the situation is one in which Lisa is doing something wrong, for example, hitting her baby brother, Jimmy, for no reason at all. Her care giver – say, her mother – tells her that what she is doing is wrong or that she should stop hitting little Jimmy. No matter what kind of internalization technique her mother uses, we can assume that the preceding factors are constant with respect to our example.[20] If Lisa's mother uses an inductive technique, she points out to her the harmful nature and consequences of her actions, and Lisa takes note of them. In contrast, if her mother uses love withdrawal techniques – for instance, ignoring Lisa or looking at her with displeasure – Lisa experiences the anger and disapproval of her mother. And if her mother uses power assertion techniques, Lisa experiences fear of punishment or even bodily coercion. Our findings indicate that if Lisa has successfully internalized her mother's norm, the most likely sources of that success, and so of Lisa's reliable moral sensibility or background beliefs, are the repetitions of learning situations in which her mother uses inductive techniques.

I conclude that the objective facts of the action play a causal role in inductively based learning that they do not play in power assertion or love withdrawal modes of learning (Hoffman 1975, pp. 228–39; 1988). In each of these situations, the child's and care giver's norms are the same. What varies is the mode of internalization. Inductive techniques are distinguished from power assertion and love withdrawal precisely because they involve a reference to object-side features in the situation, whereas the other two techniques do not. Assuming, then, that Lisa has been a successful moral learner, she has internalized a reliable moral sensibility that enables her to make relatively accurate judgments and form relatively correct beliefs about moral matters, in particular, matters of moral rightness and wrongness. The sources of the reliability of her moral sensibility, and thus of the background beliefs that she brings to bear in making observational judgments about moral rightness and wrongness, are the inductive techniques used by her mother. These techniques have as an essential element reference to objective moral facts about harming and helping, as well as the consequences of doing so.[21] Thus, they play a causal role both in the formation of her moral sensibility and in her perception of, and subsequent belief about, the rightness or wrongness of her own and others' actions.

To return to the stalemated dispute between the moral realist and the antirealist, I maintain, on the basis of the evidence of the causes for moral internalization, that the moral realist is justified in claiming, for instance, that the *fact that inflicting unnecessary harm on innocent people is morally wrong* plays a part in Lisa's belief that it is wrong. Appeal to merely the psychological and/or sociological sources for that belief is insufficient. Thus, I conclude that current empirical findings support the objectivist moral realism urged by moral realists rather than the subjectivist or intersubjectivist position supported by antirealists.

One objection to the use of these studies to support moral realism is that the studies are correlational in nature. It is fallacious to argue for any causal conclusions on the basis of correlational data. However, there are good reasons to believe that if any causal relations are involved at all, the line of causality is from the discipline technique to the state of moral internalization rather than vice versa (Hoffman 1975, pp. 228–39). First, the use of disciplinary techniques to a large extent precedes the presence of any internalization. In addition, since children generally conform to the demands of their parents, whether these demands be in the moral or the nonmoral realm, the likelihood that the direction of causality is from the technique to the state of the child is much higher than that of the reverse. Indeed, it has been estimated that young children experience pressures from parents to change their behavior, on average, every six to seven minutes during their waking hours, and for the most part they comply (Hoffman 1983, pp. 245–46). Nevertheless, this criticism points up the need for studies designed to ferret out causal factors. Even if it is improbable that particular characteristics of the child determine the style of discipline used, it may be the case that some third factor is responsible for the correlation between inductive technique and successful moral internalization. Such studies should not only be macrolevel in nature, focusing on the relative influence of the major disciplinary techniques on moral internalization, they should also be microlevel. That is, the hypothesized component processes of the techniques should be tested (Hoffman 1988, pp. 497–548).

However, even if one grants a realist interpretation of the empirical findings, it does not follow that the features of these moral facts are clearly, let alone completely, specified by these empirical findings. Indeed, there is no reason for a moral realist to make such a claim. Ethical theories are required for the specification of the nature of moral facts. But what the findings do indicate is that moral facts in an objective sense of that term play a role in the explanation of moral beliefs, and because of that, those explanations are more adequate than explanations that do not make such appeals.

Finally, since the thesis of moral realism I am advocating is an empirically based position, it could turn out that it is false and that the antirealists are, in fact, correct after all. My claim is only that current psychological findings about moral internalization favor a realist rather than an antirealist thesis. Indeed, even if moral realism turns out to be the empirically superior view, the moral realism involved need not be a commonsense moral realism. The moral properties or facts claimed by the moral realist to play a causal role in the explanation of moral and nonmoral phenomena need not be those identified by means of our commonsense accounts. The nature of these moral facts may turn out to be more accurately characterized by future scientific theories.

Thus, I conclude that there are good reasons for supporting moral realism in the strongly emergent sense of Thesis 8. On that basis, I infer the truth of Thesis 1 about the existence of moral facts. In addition, of course, the truth of the claims about moral agency and the objective moral phenomena provide support for Theses 2, 3, and 4. So, the realist integrationist contends that there is good reason to support the realist program based in her realist theses.[22]

9.5 MORAL REALISM AND THE INTEGRATIONIST'S SCIENTIFIC NATURALISTIC ACCOUNT OF MORAL AGENCY

We have been concerned with a scientific naturalistic account of moral agency. Having established that the integrationist can provide solid answers to the questions of acquisition, action, and relevance concerning moral agency, we addressed the question of adequacy. That question, of course, involved us in central issues of epistemology, particularly those of justification and of showing how the integrationist can provide a satisfactory scientific naturalistic account of the justification of moral claims. But questions and claims about knowledge raise questions about truth, and the latter about reality. So, in this chapter, we have found ourselves in the metaphysical woods. We have now emerged − I hope with a feeling of satisfaction − proclaiming a doctrine of moral realism. Indeed, after our epistemological and metaphysical exertions, we may be hoping that we will soon arrive at our destination and enjoy a well-deserved rest. It might be good, then, to see exactly where we are in light of the integrationist's original ambitions about proposing a plausible naturalistic account of moral agency.

To do this, let us return for a moment to the four substantive hypotheses that the integrationist originally proposed in Chapter 1 as constitutive of her proposal about the connections between biology, psychology, and moral agency. It is these hypotheses, we remember, that are at the heart of her biologically and psychologically based account of moral agency. Reflecting on these hypotheses, we can see that we have devoted almost all of our attention thus far to establishing the explanatory, critical, and metaethical hypotheses. By working out and providing a basis for my account of a biologically and psychologically based moral agency, I have provided support for the explanatory hypothesis, and thereby answered three of the central questions of moral agency, those of relevance, acquisition, and action. But in doing so I have, in addition, addressed the critical connection hypothesis by showing how a scientifically based naturalistic account of moral agency can retain, in a modified fashion, aspects of our folk psychological account of agency that seem to be necessary for moral agency while at the same time giving plausible responses to the challenges of reductionist or eliminativist positions about human agency. I claim, then, to have provided solid support for the explanatory and critical connection hypotheses. I also claim to have made good progress toward establishing the metaethical hypothesis. The scientific naturalistic account of justification provides a plausible way to answer the question of adequacy, and the metaphysical implications of that account allows one to explain the nature and function of moral agency in terms of a realistic account of the subjective, intersubjective, and objective realities that constitute moral reality. So, our expedition is well on its way to completion; but it still has a way to go. In terms of my hypotheses, I have yet to address the meaningfulness hypotheses. It is time to do so.[23]

NOTES

1. I use the term *facts* to refer to things, events, kinds, properties, etc. Its use is not meant to commit us to any particular ontology.
2. A classical error theorist like Mackie (1977) contends, in opposition to the noncognitivists, that moral beliefs can be true but that, in fact, none are.
3. We have seen that there are two other major theories of truth besides correspondence and warranted assertability: coherence theories and pragmatic theories. Considered as theories of truth and not merely of justification, they could be considered antirealist. I do not explore the antirealist or realist implications of these theories of truth.

4. According to constructivism, a belief is true if it conforms with certain standards of evidence or justification. For instance, according to Charles Saunders Peirce's (1955) famous formulation, a belief is true if it is the final belief of a group of ideal investigators in the end times. In other words, constructivists do not distinguish as sharply between truth and justification as do correspondence theorists. For constructivists, the truth of a belief is not constituted by an independent reality to which the belief must somehow correspond and the justification for which is indicative of such correspondence. Rather, the truth of a belief is constituted by the (right sort) of evidence for it or by its warranted assertability.

5. I shall add one more element to integrationist moral realism later.

6. It might be more appropriate to use the term *theological facts* or *divine facts* since in some Christian theological discourse *supernatural* has a more technical connotation, referring to those aspects of the divine–human encounter, such as sanctifying grace, that are completely beyond the capacities of human effort to acquire or merit. Such realities are distinct from, for instance, the creation and preservation in existence of creatures by the divine creator. However, since *supernatural* is generally used in discussions of moral realism in this generic fashion, I follow ordinary usage on the matter.

7. Besides their allegedly nonempirical character, explanations in terms of final causes have been charged with two other fatal flaws. First, they pose a form of causality that is either intrinsically incoherent or at a minimum physically impossible, since they assume that some future event is causally influential in bringing about some antecedent event. They thereby attribute causal efficacy to realities before they even exist. Second, final-cause explanations are said to be fatally anthropomorphic. Many philosophers have shown that these charges are not fatal to all explanations in terms of final causes. Wright's (1976) seminal work has been widely discussed, critiqued, and refined but remains in its basic form a widely accepted way of understanding teleological, goal-directed, or functional explanations. For a recent positive assessment see Salmon (Kitcher and Salmon 1989), and for a critical review see Schaffner (1993). Moreover, we have already seen that the charge of anthropomorphism, for instance, with respect to the theory of natural selection, can be avoided. We shall see that the charge of incoherence does not apply to the final cause explanations that are important in the integrationist's account of moral agency.

8. I shall focus on acquisition and on "ultimate" explanations. One could do similar analyses of proximate explanations of acquisition, as well as of explanations of activation.

9. We need to keep distinct two different senses of *objective* and so two different questions about the objectivity of values: the question of whether values are objective, in the sense of having an appropriate sort of independence from beliefs and desires, and the question of whether values that have that sort of independence can then refer only to the properties of nonsubjective entities like, for instance, the Platonic forms or Moore's good. It is the former question, not the latter, that I am concerned with at this point; it is, in fact, another way of asking

the realism question. One can be realist or antirealist with respect to subjects, groups of subjects, or objects. The integrationist realist is realist about all three.

10. Many philosophers have come to reject the distinction between analytic and synthetic statements. The rejection of this distinction has led to the questioning of the existence of a priori claims and knowledge. Kant distinguished between analytic a priori and synthetic a priori claims. Examples of the latter were to be found, for instance, in Euclidean geometry, arithmetic, and the fundamental laws of Newtonian physics. Kant believed, for instance, that the axioms of Euclidean geometry added to our knowledge of geometry; that is, that the predicates of those claims were not contained in the subjects, as would be the case if the claims were analytic. Moreover, Kant claimed that these claims were justified in a nonempirical fashion. Thus they were a priori. With the coming of non-Euclidean geometries, and for other reasons, many philosophers have thought it necessary to deny the existence of synthetic a priori claims. Thus, the a priori has come to be identified with the analytic, although the former is primarily and originally concerned with an alleged mode of knowledge and a sort of nonempirical mode of justification, and the latter with a semantic claim about the relationship between the content of the subject and predicate terms of a sentence. With the linguistic turn in Anglo-American philosophy, analytic claims are seen to be justified on the basis of linguistic convention and logically correct meaning derivations. Of course, competent linguistic usage may have an empirical component.

11. The reason for invoking the semantic test for properties seems to be that since characterization of properties invokes necessity, necessity has been considered in the empiricist tradition inspired by Hume to be a property of language rather than of things (Brink 1989). Thus necessity is associated with the analytic. And, of course, the analytic is a matter for the a priori.

12. Thus, for instance, (1) the moral rightness or wrongness of an action might explain why one comes to believe that the act is right or wrong; (2) the moral virtue or vice of a person might help explain judgments about the moral character of that person; (3) the moral rightness or wrongness of practices or institutions might explain the moral beliefs and the reactions people have to these practices and institutions.

13. I say "relatively" sufficient because, of course, many other factors are needed to explain helping behavior, for instance, neurophysiological processing in their respective brains. But the necessity of such factors is not in question. Harman (1986) does not allow that (M) is a testable *explanatory* hypothesis unless it is accompanied by a moral theory that contains a naturalistic *reductionistic hypothesis* – for example, that moral wrongness and rightness are constituted by certain beliefs along with pro and con attitudes.

14. This is true, of course, only if there is no causal overdetermination, that is, there are no other sufficient causes for the event to be explained. If there are other such causes, then even if the putative explanans in question had not occurred, the explanandum may have.

15. Indeed, Brink suggests that in order to explain moral judgments and beliefs, we may have to appeal to the moral learning of the individual involved. Moreover, he claims, "unless we regard this education as mistaken or in some way biased, its explanation is likely to refer eventually and in part, to moral facts, their recognition and transmission" (Brink 1989, p. 186). This is a fruitful intuition and one whose accuracy can be pursued by an investigation of the relevant empirical literature.

16. Psychoanalytic theories allot a large role in moral learning to the parental discipline technique of love withdrawal, as well as occasions for identification, whereas behaviorists and social learning theorists have traditionally emphasized the role of power techniques and occasions for imitation. For psychoanalytically oriented theorists, identification has a large dose of mentalistic mechanisms, both conscious and especially unconscious, connected with Oedipal and Electra complexes. Social learning theorists and behaviorists, who eschew mental baggage, prefer imitation, since it can be construed without that kind of baggage. Moral cognitive developmental theorists have emphasized, as situations of moral internalization, occasions of moral dissonance such as the classical moral dilemmas proposed by Kohlberg.

17. Although it is something of an overgeneralization, behavioristically oriented social learning theorists predict the effectiveness of power techniques and psychoanalytic theorists the effectiveness of love withdrawal techniques.

18. Inductive techniques also serve to give children information pertaining to the cognitive component of the moral norm to consider others. Aside from indicating the harmful consequences of their actions, induction may also communicate a prohibition against harming another, the reasons why particular actions are right or wrong, and the parents' values in regard to considering others. The association of the moral norm with both empathic feelings, particularly empathic distress, and guilt feelings makes it a *hot cognition,* one that has motivational power. It can thus enter into future considerations as a motivator independently of any concerns about approval or disapproval or fear of punishment. Some operational measures of internalization used in studies of moral internalization are (1) the amount of resistance offered by the child to pressures and temptations to behave contrary to the standard; (2) the extent to which the actions of compliance are performed without thought of sanctions for failure to comply; (3) the amount of guilt experienced/expressed by the child following failures to comply with the standard; and (4) the tendency to confess and accept responsibility for one's deviant behavior.

19. There is a positive correlation between love withdrawal and the inhibition of anger on the part of the child.

20. If the results are robust, and they appear to be, then this is true generally. The only relevant difference in the learning situation is the internalization technique.

21. It may be objected that the care giver could refer to any sorts of objective facts in making use of inductive techniques. Consequently, objective moral facts are, in fact, irrelevant to the successful learning. This sort of situation cannot be ruled

out a priori. Nor is the claim that Lisa is a completely successful moral learner or that her mother is without moral error or ignorance. So, Lisa could have learned some things she ought not to have learned. On the other hand, to make her case, the objector needs to do more than propose hypothetical counterexamples; she needs findings that match the strength of the findings we have discussed. In addition, as we have noted, inductive techniques work in part because, according to Hoffman, they engage built-in empathic capacities. We have seen that there is substantial evidence that these capacities are engaged without moral training by the objective distress of others.

22. Besides Harman's argument from explanatory inertness, there are three important classic objections to moral realism: (1) the argument from moral diversity and disagreement; (2) the argument from the odd nature of would-be moral facts, i.e., their prescriptivity; and (3) the argument from the naturalistic fallacy. I contend that I have refuted the argument from the naturalistic fallacy and Harman's argument from explanatory inertness. Although I cannot develop responses here, I also maintain that what I have said about good-consequence explanations and teleological causality goes a long way toward meeting the argument from oddness. I think that my comments on the relativity of justifications are part of an adequate reply to the argument from diversity and disagreement. For more on the latter, see Brink (1989).

23. There is another significant metaethical issue to explore. I have discussed both the justification of moral claims and the moral realities to which these claims refer. These issues are metaethical because they do not deal directly with moral issues, for instance, the questions about the morality of capital punishment or economic justice. Rather, they are concerned with the kind of knowing and the nature of the reality that is involved in the exploration of and answering of moral questions. Thus, one further metaethical question that we should address is concerned with the nature of the discipline that addresses these sorts of normative questions and attempts to find answers to them. Our examination in the last two chapters of the epistemological and metaphysical status of moral claims provides us with a way to answer the question about the nature of ethics as discipline. It should come as no surprise that the integrationist will argue that it is time that moral philosophers follow the lead of their epistemological colleagues and begin in earnest the process of naturalizing ethics. But, unfortunately, I shall have to leave that task for another day.

V

Integrating a personalistic and naturalistic view of moral agency

10

The manifest and scientific images of morality

How can we integrate our ordinary and scientifically based views of moral agency?

In Part IV, I presented an integrationist scientific naturalistic account of moral agency. In doing so, I provided support for the critical connection hypothesis and the metaethical connection hypothesis. In the preceding parts, I attempted to make the case for the explanatory connection hypothesis. At this point, then, I contend that a plausible case has been made for three of the four hypotheses that I set out to establish. I now turn to establishing the remaining integrationist hypothesis: the meaningfulness hypothesis.

Questions of meaningfulness are intimately connected with how one views one's life and the living of one's life. The particular challenge that this question poses for an integrationist is to show how a scientific naturalistic account of moral agency can make any meaningful connections with the life of ordinary moral agents. In Section 10.1, I begin to answer that question by making use of the notions of the scientific and manifest image of humans developed by Wilfrid Sellars (1963) and argue, as did Sellars, for a synoptic vision of the two in terms of the notion of the human person. This will return us in Section 10.2 to the reductionist predicament discussed in Chapter 7. Although I offer a tentative resolution of that predicament making use of a synoptic vision of human moral agents as persons, my resolution by no means settles all the issues that the predicament raises; nevertheless, it lends further confirmation to the integrationist's critical connection hypothesis and provides the entry way into an examination of the meaningfulness hypothesis and the integrationist's case for it in Section 10.3. Finally, in Section 10.4, I take stock of what we have done and what is left to be done.

Personalistic and naturalistic moral agency

10.1 THE MANIFEST AND SCIENTIFIC IMAGES OF MORALITY

In a classic essay, Wilfrid Sellars (1963, pp. 1–40) posed and offered a solution to the problem that in some ways has haunted our entire effort to present a scientific naturalistic account of moral agency and the moral life, that is, the question of the relationships between a scientific account of morality and what appears to be a distinct account of morality, an ordinary, everyday account that I have often dubbed, following current philosophical fashion, folk psychological. Sellars's primary concern is with the nature and role of philosophy. Central to his conception of philosophy is its integrative function. Specifically, Sellars argues that the job of contemporary philosophers is to show how two very different but complete images of humans are to be related. The major options are ones that are familiar to us already: (1) reduction of the ordinary, everyday image to the scientific, (2) elimination of one or the other of the images, or (3) integration of the two.

Although Sellars contrasts the ordinary, everyday account of humans, what he calls the *manifest image*, with a scientific image of humans, he does not intend to say that the manifest image is a nonscientific image. The manifest image shows signs of having been articulated through the use of some parts of scientific method, both in terms of its conceptual development and in terms of its empirical support.[1] On the other hand, when Sellars speaks of the scientific image, he is not contending that we now have one complete scientific account of humans. Moreover, both images are idealizations that a philosopher uses to highlight certain substantive and methodological differences between the two accounts of humans. Of particular importance for our consideration is the substantive difference between humans conceived of as persons in the manifest image, that is, as agents who act by means of intentions, and reality as conceived of in the current and projected scientific image, which seems to be devoid of intentional agency. Equally important is a methodological difference between the two images. Although the manifest image does make use of the scientific method, it confines itself to the formation of explanatory and predictive patterns cast in concepts that give privilege to observationally based categories, those in terms of which we describe physical and living things, animals and humans. On the other hand, the concepts used in the scientific image are theoretical. Although the existence of a distinction between theory and observation and the nature of that distinction are controversial – due in some noninsignificant part, indeed, to the pioneering work of Sellars himself – nevertheless, the *subject–matter* distinction that Sellars has in mind here is arguably a tenable one. Some aspects of reality are not accessible to us directly by the use of our unaided perceptual capacities. Some things are too small and others too large; some processes occur too quickly and others too slowly. We can directly see only the visible part of the electromagnetic spectrum. Some sounds are too high-pitched and others too low-pitched for us to hear. Thus, our senses are tuned to the middle-sized world, and micro- and macrocosmic things and processes are not directly open to our unaided perceptual capacities. Thus, theoretical terms are those that describe aspects of micro- and macroreality that are not available to us by means of our unaided sensory powers. It is in these latter terms that humans are described in the scientific image.

What image, then, are we to use in viewing humans? Sellars contends that a reduction of the manifest image to the scientific image is not generally feasible. On the other hand, the elimination of one image or the other is also unsatisfactory. An instrumentalist interpretation of the scientific image eliminates the ontological claims of that image and makes of it a merely predictive device. Thus, it gives pride of place to the ontology of the manifest image. But Sellars argues that such instrumentalism does not do justice to the cognitive achievements of the sciences and so must be rejected. On the other hand, an elimination of the manifest image poses serious problems. This is particularly so since Sellars conceives of humans precisely as the sorts of beings that conceive of themselves in terms of the manifest image. Sellars holds both that "science is the measure of what is and what is not" and that "if man had a radically different conception of himself he would be a radically different kind of man," for "man is the being which conceives of itself in terms of this [manifest] image" (1963, pp. 6, 15). The elimination of the manifest image seems to imply conceptual, epistemic, and metaphysical self-annihilation! Thus, Sellars opts for an integration of the images. His goal is to formulate a synoptic vision of human beings in the world, a scientific vision that does justice to our ordinary view of ourselves as cognitive and intentional agents.

It should be clear that the integrationist shares Sellars's problem. Although she focuses her efforts more narrowly, seeking to provide a scientifically based account of moral agency, she finds herself agreeing with Sellars's scientific realism and the ontological primacy that it allots to the scientific image of moral agency. On the other hand, she herself has conceded the importance of folk psychological accounts of moral agency and has argued that the social cognitive account of moral agency that she supports does not necessarily abandon folk psychological concepts. Thus, in some ways, she seems to be even more committed to the manifest image than is Sellars. So, let us turn briefly to Sellars's solution to the problem of the relationship between the manifest and scientific images, that is, to his

257

account of the synoptic vision. We can then examine to what extent the integrationist might want to make use of Sellars's solution.

Sellars finds in the manifest image of humans three important features that need to be integrated into the scientific image: humans' sensory, conceptual, and intentional capacities. After suggesting a way in which the first two can be fitted into the scientific image, Sellars turns to our intentional capacity.[2] It is this capacity that is of particular interest to us. According to Sellars, to consider humans as intentional beings is to consider humans as persons, that is, to think of them as beings subject to norms and to rights and duties of various sorts. Sellars contends that norms are intrinsically communitarian. Thus, to think of humans as persons is to think of them as members of communities. But what gives unity to a group, what makes it a community, are shared intentions. These shared intentions provide the norms and standards that guide the individual members of the community in the formation of their intentions. But intentions are not the sort of reality that is captured merely by descriptions, by saying what is the case or by explaining how or why something is the case; something more than a description is required. Rather, intentions are referred to by being enunciated. As Sellars puts it, "to recognize a featherless biped or dolphin or Martian as a person requires that one think thoughts of the form 'We (one) shall do (or abstain form doing) actions of kind A in circumstances of kind A.' To think thoughts of this kind is not to classify or explain, but to rehearse an intention" (1963, pp. 39–40). "And it is this something more which is the irreducible core of the person" (1963, p. 39).

Because this something more is irreducible, the manifest image of humans as persons cannot be identified with features of the scientific image, nor is it eliminable. But, on the other hand, because it is not in its distinctive aspect a descriptive category, its retention does not imply that the integration of the manifest and scientific images is dualistic. Humans as persons can be "joined to" the scientific image rather than "reconciled with" it precisely because to speak of humans as persons is not to say more about what is the case but to rehearse what ought to be the case.

What does Sellars have in mind here? Why is it that such a rehearsal does not involve the descriptive categories of the manifest image? It is because the rehearsal can be made in terms of the *scientific image*. As persons we can appropriate the world of the scientific image by speaking of the actions that we intend to perform and the circumstances in which we intend to do them, as well as the norms that guide these actions, by means of the categories of the scientific image. In other words, we can understand Sellars to be claiming that just as we can use theoretical descriptive vocabulary in our immediate perceptual reports about the world around us, so too can we use the theoretical descriptive vocabulary of the scientific image of humans to enunciate what we intend to do and the normative bases of these intentions (Rottschaefer 1982b, 1987). As Sellars (1963, p. 40) puts it:

> Thus the conceptual framework of person is not something that needs to be *reconciled with* the scientific image, but rather something to be *joined* to it. Thus to complete the scientific image we need to enrich it *not* with more ways of saying what is the case, but with the language of community and individual intentions, so that by construing the actions we intend to do and the circumstances in which we intend them in scientific terms, we *directly* relate the world as conceived by scientific theory to our purposes, and make it *our* world and no longer an alien appendage to the world in which we are living. We can, of course, as matters now stand, realize this direct incorporation of the scientific image into our life only in imagination. But to do so is, if only in imagination, to transcend the dualism of the manifest and scientific images of man-of-the-world.

It is clear that the integrationist will find much that is helpful in Sellars's synoptic vision of the manifest and scientific images of humans. Although moral norms and intentions are not the only sorts of norms and intentions that Sellars includes in the notion of humans as persons, clearly moral agency is a central feature of humans as persons. By means of their moral agency, humans are members of a moral community of persons. Moreover, as we have seen, although the most plausible current account of moral agency is, in the integrationist's view, one that at present still seems to involve some folk psychological conceptions of agency, it is not essential to her picture that any folk psychological descriptive categories of moral agency remain in the scientific end times. So, with Sellars, she can envision in imagination, at least, a time in which humans describe their agential capacities in the theoretical vocabulary of some future biological and psychological science, without the use of any folk psychological descriptions. Nevertheless, humans as persons, as particular moral agents, have been neither eliminated nor retained in some dualistic fashion. They have been joined, as Sellars puts it, to the scientific image. For whether conceived in folk psychological terms or in terms of the biological and psychological theories of the scientific eschaton, humans remain moral agents because, in fact, they so intend and act.

This Sellarsian solution may seem to be too good to be true, even for the integrationist, whose confrontation with the challenge of the reductionist predicament we have recounted in Chapter 7. The proponents of that predicament, to put it succinctly, contend that the integrationist has to choose between a nonscientific account of moral agency and a scientific account of moral agency that eviscerates it. So, before we can leave the issue of how the integrationist relates our practical and moral life with a theoretically described scientific world, it may be helpful to consider that predicament once more in light of the proposal that the integrationist adopt, as a way of linking the moral life with the theoretical scientific realm, a Sellarsian-inspired synoptic vision of the unity of the manifest and scientific images of moral agency.

10.2 THE REDUCTIONIST PREDICAMENT REVISITED

The heart of the reductionist and eliminativist challenge to the integrationist is the contention that to the extent that she uses social cognitive theories of human agency to understand moral agency, these theories must remain on the folk psychological level. But to the extent that psychologists and neuro-scientists develop adequate theories of human agency, our understanding of agency will move away from folk psychological categories. In the last analysis, the integrationist seems to be at cross purposes. Integrationism, in the final analysis, must be rejected for either separatism or antagonistic interactionism.

Does the synoptic vision of moral agency reveal this threat of fundamental incoherence to be merely that? There is good reason to think so. For the rendition of the predicament we have been discussing has as a presupposition the elimination of *agency,* including normatively guided agency. But, as we have seen, Sellars finds in agency something more than descriptive and explanatory accounts of what is the case. He finds statements of intentions and of what ought to be the case, as well as actions based on normatively guided intentions. Thus, even though who and what we are as humans turns out to be correctly described in the theoretical terms of some disciplines of the scientific eschaton, we remain agents, and agents in a world so described because it is of that sort. This is so not merely because we are able to understand and state our agency in terms that accurately describe them in this manner, but most of all because we can speak our intentions to ourselves and others in those terms and so act on them. For that reason, we remain persons, in particular moral agents, even within the scientific image of things.

But the foes of integration may contend that this solution is a trivial one. It merely says that the world stays the way it is. Humans have been, are, and will continue to be moral agents. Descriptions of that agency change, but humans go on being moral agents. In response, the integrationist insists that the issue is not merely whether we can describe our agency in different ways; the question is whether such descriptions are accurate or not. Her opponents will, no doubt, reply that if our accounts of moral agency are inaccurate, indeed false because there is no such thing as moral agency, it does no good to continue to say that there are moral agents, even if these supposed moral agents rehearse in the language of the scientific eschaton what they are about to do, the circumstances in which they are going to do them, and the rules according to which they are going to do them. The integrationist resists this new onslaught, contending that it begs the question by assuming that folk psychological and other prescientific-eschaton accounts of moral agency are radically false in the sense that they refer to nothing. It begs the question to suppose that moral agency is like demons rather than like matter. Our conceptions of the latter have changed radically in the course of scientific investigation, but from a scientific naturalist perspective there has been referential continuity as our understandings of matter have deepened through scientific investigation. On the other hand, this cannot be said about demons.

The integrationist's foes may reply that she has saved her synoptic vision by envisioning a wimpy sort of eliminativism: We have really been talking about *something* all along; it's just that we have completely misdescribed it! But whether the integrationist's projection is wimpy or not will depend on the way things are, granting that we will be able to find out how they are. Moreover, such projections are empirically based matters; and, on that score, the integrationist contends that she is on what currently appears to be some fairly stable scientific terra firma. What she has in mind is that the current scientific image of things seems to have a place both for causality and for levels of organization. And, if her position on strong emergence is correct, then systems at higher levels of organization, as well as systems of entities at lower levels of organization, exercise causality. Since humans are one sort of system at a higher level of organization, it is plausible to conclude that human agency has been retained within the current scientific image of things. However, we have heard this sort of conversation before, and since the future is not yet with us, it seems that both the integrationist and her foes will have to be satisfied with the hope that some future disciples will have a better sense of the fate of the synoptic vision.

But there is one more issue that we should address briefly before we move on to the meaningfulness hypothesis and to the question of whether the integrationist's account of things provides the sort of meaning for the moral life that we ordinary folk who think of things most frequently, if not always, in folk psychological terms are accustomed to. We can link this final consideration to the critical connection hypothesis. We have seen that the integrationist believes that a scientific naturalistic approach requires the correction of less adequate religious, a priori philosophical and folk psychological views of moral agency, and that indeed, her account of moral agency provides for such corrections. But now we turn to a much more radical conception of correction, one that goes to the very heart of who we are. Indeed, this remaining issue has to do with how we normally conceive of ourselves and how these changes in self-understanding wrought by scientific advances affect us. Once again Sellars provides us with our starting point.

We recall that Sellars claims, on the one hand, that we are the sorts of beings that we are because we conceive of ourselves in terms of the manifest image and that if we change our conception of ourselves radically, we will become radically different sorts of beings. On the other hand, Sellars is a committed scientific realist who contends that we find out what is the case about things, including ourselves, through our best scientific knowledge. Consequently, it seems that the very process of discovering scientifically who we are leads to the remaking of ourselves. Or, more subjectively, we might conclude that since we view ourselves differently in the scientific and manifest images, we are different.

How exactly does a scientific naturalistic account of moral agency affect us? Obviously, any new knowledge, indeed any new beliefs, change us by the mere fact of our having them. But, more deeply, new knowledge or new beliefs may lead to actions that change us individually, socially, or both. But what about the fundamental sort of conceptual change that seems to be involved in moving from a manifest image (or folk psychological account) of moral agency to a scientific one? Does this sort of conceptual revolution itself produce an ontological transformation? The answer to this question depends in part on the extent of that revolution. For the moment, let's focus on the conceptual revolution and set aside the practical implementation envisioned in achieving a synoptic vision. How deep might the reconceptualizing of our self-understanding as moral agents go? To answer this question, we need to know how deeply the folk psychological framework is embedded in our cognitive self-identity. Are we the sorts of cognitive systems that, because of our genetic makeup, view things about our cognitive life in folk psychological terms? If so, then nothing short of evolutionary change or genetic engineering will be sufficient to eliminate our folk psychological understanding of ourselves and our moral agency. Of course, that does not mean that we will not be able to realize the inadequacy of such conceptions, if indeed they prove to be inadequate, for understanding moral agency. For example, we continue to talk about the sun rising and setting, even though we know better. We may know that Newtonian physics is superior to Aristotelian physics and live our cognitive lives accordingly, even though we can, and sometimes do, revert to the Aristotelian perspective and categories. Or, finally, we may be able to make our immediate perceptual reports in theoretical terms, even though it is possible to retrieve the perceptual categories on which such theory-laden observations are built. Thus, if our folk psychological understanding of moral agency is part of our evolutionary heritage, it may be ineliminable, but nevertheless fully subject to critique and to epistemic and ontological replacement.[3] On the other hand, our folk psychological conceptions may not be so fully entrenched in our genetic makeup; they may be social/cultural acquisitions. If so, changes to a scientific conception of moral agency may be more easily made and folk psychological conceptions may fall by the wayside like those of phlogiston and the heavenly spheres. Our integrationist need not take a stance on this empirical question. But the embeddedness of the folk psychological account of moral agency in our genetic makeup, as well as the degree to which that conception must yield to a scientific account of moral agency because of the latter's epistemic superiority, are factors that affect the nature of the critical connection hypothesis as it applies to the question of how extensively we will have to modify our conceptions of ourselves as moral agents, given the scientific naturalistic perspective of the integrationist. Finally, the practical changes envisioned in a synoptic union of manifest and scientific images are in part dependent on the degree of conceptual correction that is required in accepting a scientific conception of our moral agency. These changes themselves will have further effects. To the extent that an account of moral agency can be applied as a kind of personal science, the integrationist will in general view these changes as positive, morally significant, and progressive.

I now rest my case for the critical connection hypothesis and, in addition, contend that I have laid the foundations for the support of the meaningfulness hypothesis. Nevertheless, there are still questions about the meaningfulness of such a moral life. Is there really any basis for an optimistic assessment of moral life conceived and lived within the scientific image? What does it matter that we can integrate the manifest and scientific images of moral agency and ethics into a synoptic vision if it doesn't have any meaning for us? It is time to address directly the meaningfulness hypothesis.

10.3 CAN AN INTEGRATIONIST ACCOUNT OF MORAL AGENCY REALLY BE MEANINGFUL TO ANYONE?

I have been concerned with issues in metaethics, the subdiscipline of ethics that deals with the nature, function, justification, and ontology of morality.

263

A very large part of our inquiry has focused on the nature of morality, conceived of in terms of moral agency. But what about the function of morality? What is it for? That question might seem to be an easy one to answer. Moral agency and morality are for attaining morally good ends, for doing the morally right and avoiding the morally wrong. Yes, of course; but the question of function has more to it than that. Why is it important to do what is morally right and avoid what is morally wrong? This question is not the same as the question Why be moral rather than nonmoral or immoral? I can be completely convinced that I ought to be moral, that is, that I ought to develop my moral agency and do what is morally right, but still have questions about the function of morality. But, you might ask, isn't the morally good intrinsically valuable? Isn't it worth seeking for its own sake? Morality should be its own reward. So, the function of morality and moral agency is to achieve the moral good. But even if we concede that point, the question of the function of morality still seems to go deeper. What more could there be?

I raised the question of the function of morality earlier in terms of what we have called the meaningfulness hypothesis. The integrationist sees it as part of her task to show that an evolutionary view of human life and a psychologically based view of social and personal life make a moral life meaningful. So, another way of putting the traditional question about the function of morality in the context of our inquiry is to ask how significant morality is from the scientific naturalistic perspective of the integrationist. After all, if her account of moral agency and her view of ethics as scientifically based diminishes or evacuates the meaningfulness of morality, then perhaps it is fatally flawed, and we ought to reconsider our acceptance of it.

I argue that the integrationist view provides some partial meaning to our moral lives, but that that meaning is limited because the sciences of biology and psychology are themselves restricted in what they tell us about human existence and the world in which humans live. Thus I contend that from a scientific naturalistic perspective, the other natural sciences, as well as the social sciences, are needed to help fill in the picture. Nevertheless, I argue that a good case can be made for the meaningfulness of our moral lives conceived from a scientific naturalistic perspective. Consequently, I claim that I have provided some substantial support for the last of the four integrationist hypotheses, the meaningfulness hypothesis.

An antagonist opponent of integrationism argues that meaningfulness is a function of self-originating projects; and since the integrationist's conception of morality makes of it a science, it is the antithesis of such projects. The integrationist can resist this charge on the basis that some values are not entirely self-originating, for example, such values as adequate food, clothing, shelter, and social relations. Moreover, she contends that both her substantive and functional accounts of moral agency provide for genuine agency in the discovery, creation, and attainment of moral values. Of course, it is not in the scientific naturalistic cards for human moral agents, as individuals, as groups, or a species, to achieve the meaningfulness of the sorts of self-originating projects that we might associate with a god. But the integrationist's opponents are more than likely willing to admit that. However, her opponents' initial gambit is helpful in that it provides us with a way to understand some of the essential elements of meaningfulness. Two of these seem to be *agency* and *worthy goals*. The integrationist contends that she has not only made genuine human agency a part of her position but also has provided solid scientific evidence for its existence. Moreover, she has argued for the existence of moral values as the objects of human moral agency. So we might wonder whether and, if so, why her opponents still find her story about the meaningfulness of a scientific naturalistic account of moral agency, morality, and the moral life less than satisfactorily meaningful.

We can get a handle on their dissatisfaction by considering some alternative accounts of meaning that might appear to provide a higher degree of meaning to our moral life than does the integrationist's. Let's consider two, a theistically based account and a humanistically based one. According to a traditional Christian account of the meaning of morality, it functions within the context of redemptive history.[4] That history starts with the creation of the universe, including our own earth and all its inhabitants, with a special plan that humans are to share in God's eternal beatific life forever. It continues with the sin of our first parents and the loss of eternal life, the incarnation of the son of God, his salvific death, and his resurrection on behalf of the sinful human race. Thenceforth human history is the working out of that salvation by individual men and women who, with God's grace, are to live a moral life by following the commandments of Jesus and the Church. Individual success will bring eternal life and failure eternal damnation, and at the end of the world, Christ will return triumphantly and the entire earth will be transformed. We can see that in this traditional Christian religious story the moral life is put into a larger context, the context of cosmic history. Moreover, and even more important, that history has its origins in the beneficent intentions of an all-loving and all-powerful creator, and it is guided by the active providence of that creator, along with the salvific work of his divine son and the Holy Spirit, who ensure that its ending will be a blessed one. Within this narrative, living a moral life has a larger meaning than that of doing one's duty and achieving the moral good because the human and created good are taken up in a more significant and important good: that of sharing in the divine life for eternity. This is especially so for humans, who are created in the image of God and are capable of participating in his eternal life in a way not possible for other creatures, which do not possess the intellectual and volitional capacities of humans.

Let us consider, on the other hand, a nontheistic and nonscientific humanistic account of the meaningfulness of the moral life. Although some humanistic accounts might confine the story of the moral life to humans and to the short lives of individuals, there is no intrinsic reason why they need to do so. Thus, even though they may not have the cosmic sweep of the theistic story, the narratives in which the moral lives of individuals are set often include human communities, the course of human history, and sometimes have a place for nature. What the humanistic story lacks is the personalization of history in terms of the plans, intentions, and actions of a beneficent creator God within the created world, especially the human world. However, it seems often to have retained from that story the central place of the human species within the universe and the pivotal importance of the moral life for humans. Humans, as in the theistic conception, are set apart from nature. Possessing the freedom and dignity of autonomous agents, they are not subject in the same way to the forces that govern the rest of nature. Consequently, the significance of the moral life of humans is assured and made manifest.

These two alternative stories of the meaningfulness of the moral life for humans point to at least three apparently large lacunae in the integrationist's conception of the meaningfulness of the moral life, lacunae that can be traced, so it is alleged, to her scientific naturalist commitments. First, there is an absence of a religious element, for example, the Christian theistic narrative.[5] Second, there is no cosmic story. And, finally, the specialness of humankind is deemphasized. Are religious, cosmic, and humanistic emphases necessary for making morality meaningful? Even if they are not necessary, are they so important that without them the integrationist account of the meaningfulness of the moral life is so pallid that perhaps less adequate religious, or purely humanistically inspired, accounts of moral agency and ethics are more acceptable because of their ability to make the moral life more meaningful than can an integrationist's story? Clearly, these are very large questions that cannot be adequately addressed in the short compass that I have left. However, the integrationist owes her opponents some sense of how she might at least approach the task of exploring and answering these questions.

To start, we can identify several aspects of meaningfulness — two of which we have already noted — that seem to emerge from the theistic and humanistic accounts of the meaningfulness of the moral life that might serve as requirements for, or at least significant indicators of, meaningfulness. These are (1) genuine agency, (2) worthy goals, (3) scope, (4) coherence, and (5) harmony. Clearly, the list is neither complete nor uncontroversial. But I have tried to formulate these indicators in a theory-neutral fashion in order not to prejudice the case for one account over another. Let's grant, for the sake of discussion, that these are elements of meaningfulness that are acceptable to all parties and ask to what extent the integrationist scientific naturalistic picture of moral agency and ethics possesses these significant indicators of meaningfulness.

The integrationist contends that as far as significant agency goes, she has made a good case that humans are able to exercise genuine agency. She has provided good scientific evidence that they are able to develop and maintain an integrated set of capacities for acting and achieving their ends. The soft determinist sort of freedom that humans possess does not rid humans of all causal influences. Indeed, it is these causal influences that make possible a genuine self-determination within a reciprocally causally interacting social and natural environment. Moreover, although it has not been argued for, the integrationist contends that one's moral agency can be enhanced by means of theoretically based and effective problem-solving techniques. Thus the integrationist maintains that her account of human moral agency makes it a meaningful part of human life.

The issues of worthy goals and scope merge. The reason for this is that although there may be some major differences between religionists, humanists, and integrationists concerning substantive moral goals, there is, nevertheless, a large area of overlap. The differences focus more on the context in which the pursuit of moral goals is placed. Although the integrationist has been concerned with providing an account of the biological and psychological bases of moral agency, she is committed in principle to an examination of the role that all the natural and social sciences might play in understanding moral agency. This is not to say that even within her own bailiwick she does not have the resources for a narrative that is significant in scope. For, on the one hand, she has placed the origins of human moral agency within the sweep of evolutionary history; and, on the other hand, she can envision the extension of moral concern to include nonhuman animals, other living things, and even inorganic nature. Moreover, this scientifically based naturalistic story of the origin, development, and extension of moral agency can itself be placed in the larger picture of cosmic history provided by physicists and cosmologists. Thus, the integrationist contends that her story of the meaningfulness of moral agency concerns the pursuit of worthy goals of wide scope.

Granting the integrationist that last point, what about the mark of

267

coherence? The sources of coherence in the integrationist account derive from their bases in the patterns of interaction reflected in the regularities on which the explanatory patterns of biology and psychology are built. These explanatory patterns fit with other explanatory patterns that emerge from a study of cosmology, physics, and chemistry, on the one hand, and anthropology, sociology, economics, and political science, on the other. Given current views about quantum indeterminacy, the coherence is not entirely of the deterministic sort. Both quantum indeterminacy and emergence call for novelty, and the former for a role for radical chance. The coherence that comes from goal-directed, purposeful, and intentional activity emerges only gradually in the integrationist's story. However, in contrast with many religious views, it does not penetrate all reality, nor is it injected from the outside.

If there is some coherence in the universe, given the integrationist's scientifically based picture of it, is there any harmony? A maximal notion of harmony includes the condition of all things working toward and achieving some maximal good for all. It seems clear from the integrationist's perspective that not only is there no evidence for maximal harmony, but that there is strong evidence against it. Even if the integrationist confines her account to this universe rather than all possible universes, her conclusion remains that maximal harmony does not seem to be a feature of the way things are. Indeed, not only are coming to be and ceasing to be part of the universe as we know it, but so are conflict and destruction on all levels of reality. Ought this to be? Even if we try to answer such a question, the integrationist will remind us that any sort of ought depends on what is possible; and, as far as she can determine from what is and has been the case, maximal harmony — even within this given universe — is not possible. That is not to say that there is no evidence of systems sometimes working together even for their mutual benefit. But the evidence suggests that these systems are spatially and temporally limited in scope. At most, there seems to be evidence for only a very modest sort of harmony within our universe. Nevertheless, from the integrationist perspective, that is no reason not to foster realistic moral goals of harmony or to cease efforts to achieve them within those systems where there is some reasonable hope for success.

After this all too brief and cursory exploration of the meaningfulness of her account of moral agency and ethics, it seems clear that the integrationist cannot completely eliminate the lacunae that a comparison with humanistic and theistic religious pictures brings out. Even though the integrationist can tell a cosmic story in which humans have a distinctive evolutionary niche and specific cultural and social achievements, her story does not bring in a theistic God, nor does it set humans apart from nature. Is the integrationist account incompatible with traditional theism or theism of any sort? Is it incompatible with nontheistic religious positions of any sort? It is clearly beyond the scope of my inquiry to pursue these issues any further.[6] To what extent do humans' cognitive capacities enable them to build cultural and social systems that isolate them from the influences of nonhuman forces? To what extent are we able to control the "forces of nature" and the "forces of the cosmos," and thereby maintain and pursue our ends, whatever the nature of these forces? These too are large questions that can be raised here but left hanging.[7]

Thus, the integrationist claims that the moral life is meaningful because moral agents do possess significant agency, which they exercise with respect to worthy goals that are comprehensive in scope and relatively coherent and harmonious in form. Nevertheless, she recognizes that alternative accounts may make claims for a meaningfulness that possesses these characteristics of meaningfulness to a greater degree. Our integrationist will not dismiss these views out of hand, but will demand at a minimum that they be consistent with the best findings of the natural and social sciences. She contends that her own account not only meets the criterion of consistency, but also meets the stronger one of evidential support. In the end, she believes that the extent of meaningfulness that can be legitimately postulated by any account of moral agency must be guided by the best scientific picture available. Thus, although other accounts of moral agency may possess a greater degree of meaningfulness than her scientific naturalistic account, her account is preferable because the degree of meaningfulness that she postulates is more realistic, in the sense that it better reflects the world in both its actuality and potentiality than do alternative supernaturalistic religious or purely humanistic accounts.

But for many of her opponents, this sort of response smacks of scientism. The charge of scientism takes many forms but always has a negative connotation. Unfortunately, some who use the term do so only to characterize negatively views that they oppose. Often, scientism is considered to be the failure to see the limitations of the scientific way of knowing about reality, and the consequent ruling out as unworthy of consideration other modes of knowing that capture aspects of reality not open to science. Is the integrationist guilty of scientism, since she demands that accounts of the moral life be at least compatible with the findings of the sciences, if not supported by them? I do not think so. First, it is part of her epistemic obligation not to rule out of court points of view with which she might differ merely because she does not happen to like what they say. So, she denies that she has ruled out other potential sources of knowledge without good reason. But, second, these same epistemic obligations demand that she accept as supported only those claims that are genuinely supported. That, of course, does not mean that she accepts only those statements that she herself has determined are adequately supported, since it seems that it may not be humanly possible for one to hold only those beliefs that have passed the criteria of adequacy that one has set up for oneself (Goldman 1986; Harman 1987; Kitcher 1993). Third, she does not automatically exclude claims that there are other modes of access to reality distinct from those used by the sciences. However, she will demand that there be ways to adequately establish such claims. But doesn't such a demand prejudice the case in the integrationist's favor as soon as she makes clear what her criteria of adequacy are? Not necessarily, since the integrationist contends that we all share fundamental cognitive goals such as truth, understanding, and explanation, as well as basic cognitive capacities of perception, memory, and inference that enable us to assess in some minimally adequate way claims about the relative epistemic merits of alternative modes of knowing. In fact, she maintains that the epistemic practices of the sciences are our most reliable extensions of the use of these capacities. It should now be clear, if it was not from the beginning of this discussion, that our considerations have moved us into major issues in epistemology. But that should not surprise us, given the interconnection of philosophical issues. Although this response may not satisfy her opponents, she and they can at least agree that they have a number of issues that they still need to discuss in the future.

Where does this leave us? We have seen that the integrationist claims that she has given some support to the meaningfulness hypothesis, the claim that her scientific naturalist account of moral agency, ethics, and the moral life shows them to be meaningful. Explicating and supporting this claim has led us to touch on, without resolving, much larger issues in epistemology and the relative merits of a scientific naturalistic approach to reality.

10.4 LOOKING BACK AND LOOKING FORWARD

We have come a long way in our examination of the relationships between biology and psychology, on the one hand, and morality, on the other. In opposition to separatists and antagonists, I have adopted an integrationist position, arguing that it is preferable to positions based purely on common-sense, a priori philosophical, or independent religious sources of knowledge. My integrationist task was to set forth an account of moral agency from a scientific naturalistic perspective. I argued that a satisfactory integrationist account needed to link relevant aspects of biology and psychology in such a way as to show that these disciplines provide an account of moral

agency, ethics, and our moral life with explanatory power, a critical perspective on other approaches, epistemological and metaphysical understanding, and a significant degree of meaningfulness. I first focused on
moral agency, developing a four-tiered theory of moral agency consisting of
base, behavioral, reflective, and self-referentially reflective levels. In doing
so, I attempted to provide biologically and psychologically based answers
to the central questions of moral agency, those of relevance (what constitutes it both functionally and substantively), acquisition (what are the
evolutionary, developmental, behavioral, and social cognitive sources for
its origin, development, maintenance, and enhancement), action (how is
moral agency put to work), and adequacy (how are our moral beliefs,
motivations, and actions justified). This latter question required that I
discuss not only issues in moral epistemology but also issues in moral
ontology. From the perspective of traditional ethics, these discussions
focused on three important aspects of metaethics: the nature and function of
morality and its justification.[8] I have also argued for the meaningfulness of
moral agency and the moral life conceived of within an integrationist perspective. A central issue that arises concerning the critical connection hypothesis is the extent to which an integrationist account of morality demands its elimination or reduction to one or several of the sciences. Rather
than accepting either alternative, I argued that on the theoretical level,
current evidence supports an understanding of moral agency that retains
important links with our folk psychological conceptions. Moreover, even if
future developments in our understanding of moral agency require major
shifts in our folk psychological conceptions or their theoretical elimination,
it is not clear that such epistemic and ontological replacements require or,
indeed, can bring about the elimination of folk psychological conceptions
that may have an evolutionary basis that is eliminable only through species
change. Nevertheless, in the eventuality of even such conceptual and epistemic elimination, I adopted a Sellarsian position that maintains that our
agential capacities, grasped through folk psychological categories in the
manifest image of moral agency, can be recognized and put into play by a
moral agency conceived within the scientific image of humans as persons.

But, as we have seen in our specific, detailed discussions of the four
integrationist hypotheses concerning the relationships between the sciences
and morality, the integrationist's task is far from complete. From the scientific point of view, many of the parts of biology and psychology relevant for
understanding moral agency have only recently been developed, and much
theoretical work and empirical research are still to be done. In addition, the
scientific side of the integrationist position has been only partially laid out,
because I have confined myself to the contributions of biology and psychol-

271

ogy to the understanding and explanation of moral agency and the moral life. But, as I have mentioned several times, the social sciences of sociology, anthropology, economics, and political science will undoubtedly make important contributions to a scientific naturalistic account of moral agency and the moral life. In addition, although the scientific naturalistic perspective in philosophy may have a long and honorable tradition, its revival in Anglo-American philosophy is relatively recent. Indeed, its use in moral epistemology and ontology has scarcely begun. So much work still needs to be done on the philosophical side of the integrationist approach.

Otto Neurath's (1932) metaphor remains apt for the integrationist enterprise. Integrationists must constantly rebuild their leaky philosophical boat in transit, using whatever materials they can to keep it afloat. But the integrationists are happy about the progress made thus far and optimistic about the future course of their voyage.

NOTES

1. In what he calls the *original image* (humans' original reflective conception of themselves in nature), Sellars hypothesizes that humans thought of all reality in personal terms. Only later, through conceptual refinement, did they restrict the notion of the personal, that is, agency based on intentions, to humans and the gods, retracting its application from merely material or living entities (Rottschaefer 1982b, 1987).
2. Sellars (1968) offers a functionalist account of our conceptual capacities and argues that the continuous character of sensations can be understood within a scientific image whose basic constituents are continuous in character rather than particulate.
3. I have argued against what I take to be Sellars's position that our ordinary knowledge framework is in principle eliminable methodologically (Rottschaefer 1978).
4. Obviously, there are many other religious accounts that provide a larger context in which to view the meaning of the moral life. I select a traditional Christian one only because it will undoubtedly be familiar to many readers.
5. I do not intend to identify religion with so-called supernaturalistic religion. One may well be able to make the case for a scientific naturalistic religion, where religion is conceived of as being concerned with "ultimate reality," however that is conceived. I leave this question open and contrast the integrationist's views with those of supernaturalistic religions, in which the views about ultimate reality are formulated for the most part independently of scientific findings.
6. My inclination is to answer yes to the first question and no to the second. But this is not the place to pursue these issues.

272

7. My own tendency is to believe that the human species is significantly limited in its capacity for control and to that extent, at least, cannot set itself apart from the rest of reality. Moreover, it seems that it ought not to do so.
8. Besides metaethics, traditional ethics has a normative component and an applied component. Although I have not argued for it, from an integrationist point of view, normative ethics is best conceived as a normative science of the moral, two of whose basic subdisciplines are relevant parts of biology and psychology, while individual applied ethics is best understood as a personal science of the moral.

References

Alexander, R. D. 1979. *Darwinism and human affairs.* Seattle: Washington University Press.

1985. "A biological interpretation of moral systems." *Zygon* 20: 3–20.

1987. *The biology of moral systems.* New York: Aldine DeGruyter.

Alexrod, R. 1984. *The evolution of cooperation.* New York: Basic Books.

Alexrod, R., and W. D. Hamilton. 1981. The evolution of cooperation. *Science* 211: 1390–6.

Alston, W. P. 1967. "Emotion and feeling." In P. Edwards, ed. *Encyclopedia of Philosophy,* vol. 2. New York: Macmillan, pp. 479–86.

1971. "Comments on Kohlberg's 'From is to ought.'" In T. Mischel, ed. *Cognitive development and epistemology.* New York: Academic Press, pp. 269–84.

1977. "Self-intervention and the structure of motivation." In T. Mischel, ed. *The self: Psychological and philosophical issues.* Oxford: Basil Blackwell, pp. 65–102.

Aronfreed, J. 1968. *Conduct and conscience: The socialization of internalized control over behavior.* New York: Academic Press.

1976. "Moral development from the standpoint of a general psychological theory." In T. Lickona, ed. *Moral development and behavior: Theory, research and social issues.* New York: Holt, Rinehart & Winston, pp. 54–69.

Aune, B. 1967. *Knowledge, mind and nature: An introduction to theory of knowledge and philosophy of mind.* New York: Random House.

Bandura, A. 1974. "Behavior theory and the models of man." *American Psychologist* 28: 859–69.

1976. "Self-reinforcement: Theoretical and methodological considerations." *Behaviorism* 4: 135–55.

1977a. "Self-efficacy: Toward a unifying theory of behavioral change." *Psychological Review* 84: 191–215.

1977b. *Social learning theory.* Englewood Cliffs, NJ: Prentice-Hall.

1978a. "Reflections on self-efficacy." *Advances in Behavior Research and Therapy* 1: 237–69.

References

1978b. "The self-system in reciprocal determinism." *American Psychologist* 33: 344–58.

1982. "Self-efficacy mechanism in human agency." *American Psychologist* 37: 122–47.

1984. "Recycling misconceptions of perceived self efficacy." *Cognitive Theory and Research* 8: 231–55.

1986. *Social foundations of thought and action: A social cognitive theory.* Englewood Cliffs, NJ: Prentice-Hall.

1988. "Social cognitive theory of moral thought and action." In W. M. Kurtines and J. L. Gewirtz, eds. *Moral behavior and development: Advances in theory, research and applications,* Vol. 1, Hillsdale, NJ: Erlbaum, pp. 45–103.

1991. "Self-efficacy mechanism in physiological action and health-promoting behavior." In J. Madden, ed. *Neurobiology of learning, emotion and affect.* New York: Raven Press, pp. 229–70.

Bandura, A., C. B. Taylor, S. L. Williams, I. N. Mefford, and J. D. B. Barchas. 1985. "Catecholamine secretion as a function of perceived coping self-efficacy." *Journal of Counseling and Clinical Psychology* 51: 406–14.

Barkow, J. H., L. Cosmides, and J. Tooby. 1992. *The adapted mind: Evolutionary psychology and the generation of culture.* New York: Oxford University Press.

Barnett, M. A. 1987. "Empathy and related responses in children." In N. Eisenberg and J. Strayer, eds. *Empathy and its development.* Cambridge: Cambridge University Press, pp. 146–62.

Baron, M. 1985. "Varieties of ethics of virtues." *American Philosophical Quarterly* 22: 47–53.

Bateson, P. 1986. "Sociobiology and human politics." In S. Rose and L. Appignanesi, eds. *Science and beyond.* New York: Basil Blackwell, pp. 79–99.

1987. "Biological approaches to the study of behavioral development." *International Journal of Behavioral Development* 10: 1–22.

1989. "Does evolutionary biology contribute to ethics?" *Biology & Philosophy* 4: 287–302.

Bavelas, J. B., A. Black, C. R. Lemery, and J. Mullett. 1987. "Motor memory as primitive empathy." In N. Eisenberg and J. Strayer, eds. *Empathy and its development.* Cambridge: Cambridge University Press, pp. 317–38.

Beatty, J. 1981. "What's wrong with the received view of evolutionary theory?" In P. Asquith and R. Giere, eds. *Proceedings of the 1980 biennial meeting of the Philosophy of Science Association.* East Lansing, MI: Philosophy of Science Association, pp. 397–439.

Bechtel, W., and A. Abrahansen. 1991. *Connectionism and the mind: An introduction to parallel processing in networks.* Cambridge: Basil Blackwell.

Blasi, A. 1980. "Bridging moral cognition and moral action." *Psychological Bulletin* 88: 1–45.

1983. "Moral cognition and moral action: A theoretical perspective." *Developmental Review* 3: 178–210.

1984. "Moral identity: Its role in moral functioning." In W. Kurtines and J.

References

Gewirtz, eds. *Morality, moral behavior and moral development.* New York: Wiley, pp. 128–37.

Block, N., and G. Dworkin. 1976. *The IQ controversy.* New York: Pantheon Press.

Bloom, F. E., and A. Lazerson. 1988. *Brain, mind and behavior,* 2nd ed. New York: W. H. Freeman.

Bower, G. H., and E. R. Hilgard. 1981. *Theories of learning,* 5th ed. Englewood Cliffs, NJ: Prentice-Hall.

Boyd, R. N. 1984. "The current status of scientific realism." In J. Leplin, ed. *Scientific realism.* Berkeley: University of California Press, pp. 41–82.

Bradie, M. 1994. *The secret chain: Evolution and ethics.* Albany, NY: State University of New York Press.

Breland, K., and M. Breland. 1961. "The misbehavior of organisms." *American Psychologist* 16: 681–4.

1966. *Animal behavior.* New York: Macmillan.

Brewer, W. 1974. "There is no convincing evidence for operant or classical conditioning in adult humans." In W. Weiner and D. Palermo, eds. *Cognition and symbolic processes.* Hillsdale, NJ: Erlbaum, pp. 1–42.

Brink, D. O. 1989. *Moral realism and the foundation of ethics.* Cambridge, MA: Cambridge University Press.

Burton, R. V. 1978. "Interface between the behavioral and the cognitive-development approach to research in morality." In B. Z. Presseisen, D. Goldstein, and M. H. Appel, eds. *Topics in cognitive development.* New York: Plenum Books, pp. 115–23.

1984. "A paradox in theories and research in moral development." In W. Kurtines and J. Gewirtz, eds. *Morality, moral behavior and moral development.* New York: Wiley, pp. 193–207.

Campos, J. J., K. C. Barrett, M. E. Lamb, H. H. Goldsmith, and C. Stenbey. 1983. "Socioemotional development." In M. M. Harth and J. Campos, eds., *Infancy and developmental psychology,* Volume II of *Handbook of child psychology,* 4th ed., ed. Paul H. Mussen. New York: Wiley, pp. 783–916.

Changeux, J. 1985. *Neuronal man: The biology of mind,* trans. Lawrence Garey. New York: Oxford University Press.

Chomsky, N. 1959. "Review of verbal behavior." *Language* 35: 26–58.

Churchland, P. M. 1979. *Scientific realism and the plasticity of mind.* Cambridge: Cambridge University Press.

Churchland, P. M. 1984. *Matter and consciousness: A contemporary introduction to the philosophy of mind.* Cambridge, MA: Massachusetts Institute of Technology Press.

1988. *Matter and consciousness: A contemporary introduction to the philosophy of mind,* rev. ed. Cambridge, MA: Massachusetts Institute of Technology Press.

1989. *A neurocomputational perspective: The nature and structure of science.* Cambridge, MA: Massachusetts Institute of Technology Press.

1995. *The engine of reason, the seat of the soul: A philosophical journey into the brain.* Cambridge, MA: Massachusetts Institute of Technology Press.

Churchland, P. S. 1988. *Neurophilosophy: Toward a unified science of the mind-brain.* Cambridge, MA: Massachusetts Institute of Technology Press.

References

Damasio, A. R. 1994. *Descartes' error: Emotion, reason, and the human brain.* New York: Grosset/G. P. Putnam's Sons.

Damasio, H., T. Grabowski, R. Frank, A. M. Galaburda, and A. R. Damasio. 1994. "The return of Phineas Gage: The skull of a famous patient yields clues about the brain." *Science* 264: 1102–5.

Damon, W. 1984. "Self-understanding and moral development from childhood to adolescence." In W. Kurtines and J. Gewirtz, eds. *Morality, moral behavior, and moral development.* New York: Wiley, pp. 109–27.

Darden, L., and J. A. Cain. 1989. "Selection type theories." *Philosophy of Science* 56: 106–29.

Darwin, C. 1859. *On the origin of species.* London: Murray.

1871. *The descent of man, and selection in relation to sex.* London: Murray.

Dawkins, R. 1976. *The selfish gene.* Oxford: Oxford University Press.

Dennett, D. 1978. "Why the Law of Effect will not go away." In D. Dennett, *Brainstorms: Philosophical essays in mind and psychology.* Cambridge, MA: Bradford Books, pp. 71–89.

Dobson, K. S., ed. 1988. *Handbook of cognitive behavior therapies.* New York: Guilford Press.

Dobson, K. S., and L. Block. 1988. "Historical and philosophical bases of the cognitive behavioral therapies." In K. S. Dobson, ed. *Handbook of cognitive-behavioral therapies.* New York: Guilford Press, pp. 3–38.

Dretske, F. 1988. *Explaining behavior: Reasons in a world of causes.* Cambridge, MA: Massachusetts Institute of Technology Press.

Dulaney, D. 1974. "On the support of cognitive theory in opposition to behavior theory: A methodological problem." In W. Weimer and D. Palermo, eds. *Cognition and the symbolic processes.* Hillsdale, NJ: Erlbaum, pp. 43–56.

Dunn, J., and C. Kendrick. 1979. "Interaction between young siblings in the context of family relationships." In M. Lewis and L. A. Rosenblum, eds. *The child and the family.* New York: Plenum Books, pp. 143–68.

Eastman, C., and J. S. Marzillier. 1984a. "Continuing problems with self-efficacy theory: A reply to Bandura." *Cognitive Therapy and Research* 8: 257–62.

1984b. "Theoretical and methodological difficulties in Bandura's self-efficacy theory." *Cognitive Theory and Development* 8: 213–29.

Eckerman, C. O., J. L. Whatley, and S. L. Kutz. 1975. "Growth of social play with peers during the second year of life." *Developmental Psychology* 11: 42–9.

Edelman, G. M. 1987. *Neuronal Darwinism: The theory of neuronal group selection.* New York: Basic Books.

1992. *Bright air, brilliant fire: On the matter of mind.* New York: Basic Books.

Eisenberg, N., and P. Miller. 1987. "Empathy, sympathy, and altruism: Empirical and conceptual links." In N. Eisenberg and J. Strayer, eds. *Empathy and its development.* New York: Cambridge University Press, pp. 292–316.

Erwin, E. 1978. *Behavior therapy: Scientific, philosophical and moral foundations.* New York: Cambridge University Press.

Estes, W. K. 1971. "Reward in human learning: Theoretical issues and strategic choice points." In R. Glaser, ed. *The nature of reinforcement.* New York: Academic Press, pp. 16–36.

References

Eysenck, H. J. 1976. "The biology of morality." In T. Lickona, ed. *Moral development and behavior: Theory, research and social issues.* New York: Holt, Rinehart & Winston, pp. 108–23.

Fancher, R. E. 1973. *Psychoanalytic psychology: The development of Freud's theory.* New York: W. W. Norton.

Feldman, M. W., and R. C. Lewontin. 1975. "The heritability hang-up." *Science* 190: 1163–8.

Flanagan, O. J. 1982. "Moral structures." *Philosophy of the Social Sciences* 12: 255–70.

1984. *The science of mind.* Cambridge, MA: Massachusetts Institute of Technology Press.

Frankfurt, H. 1971. "Freedom of the will and the concept of a person." *Journal of Philosophy* 67: 5–20.

Futuyma, D. J. 1979. *Evolutionary biology.* Sunderland, MA: Senauer.

Garcia, J., and R. Koelling. 1966. "Relation of cue to consequence in avoidance learning." *Psychonomic Science* 4: 123–4.

Gardner, H. 1985. *The mind's new science: A history of the cognitive revolution.* New York: Basic Books.

Garfinkel, A. 1981. *Forms of explanation: Rethinking the questions in social theory.* New Haven, CT: Yale University Press.

Gibbs, J. C., and S. V. Schnell. 1985. "'Moral development' versus 'socialization.'" *American Psychologist* 40: 1071–80.

Giere, R. 1983. "Testing theoretical hypotheses." In J. Earman, ed., *Minnesota Studies in the Philosophy,* Vol. 10. Minneapolis: University of Minnesota Press, pp. 269–98.

Giere, R. 1984. *Understanding scientific reasoning,* 2nd ed. New York: Holt, Rinehart & Winston.

Godfrey-Smith, P. 1994. "A modern history of functions." *Nous* 28: 344–62.

Goldman, A. 1986. *Epistemology and cognition.* Cambridge, MA: Harvard University Press.

Gould, S. J. 1977. "Biological potentiality vs. biological determinism." In S. J. Gould, *Ever since Darwin: Reflections on natural history.* New York: W. W. Norton, pp. 251–9.

Gould, S. J., and R. C. Lewontin. 1979. "The spandrels of San Marco and the Panglossian paradigm." In *Proceedings of the Royal Society of London, B 205,* pp. 581–98. Reprinted in E. Sober, ed. 1984. *Conceptual issues in evolutionary biology.* Cambridge, MA: Massachusetts Institute of Technology Press.

Graham, G., and T. Horgan. 1988. "How to be realistic about folk psychology." *Philosophical Psychology* 1: 69–81.

Griffiths, P. E., and R. D. Gray. 1994. "Developmental systems and evolutionary explanation." *Journal of Philosophy* 91: 277–304.

Grimes, T. 1988. "The myth of supervenience." *Pacific Philosophical Quarterly* 69: 152–60.

Grunbaum, A. 1984. *The foundations of psychoanalysis: A philosophical critique.* Berkeley: University of California Press.

References

Hamilton, W. D. 1964a. "The genetical evolution of social behavior I." In G. C. Williams, ed., 1971. *Group selection*. Chicago: Aldine Atherton, pp. 23–43.

Hamilton, W. D. 1964b. "The genetical evolution of social behavior II." In G. C. Williams, ed., 1971. *Group selection*. Chicago: Aldine Atherton, pp. 44–89.

Harman, G. 1977. *The nature of morality*. New York: Oxford University Press.

1985. "Is there a single true morality?" In D. Copp and D. Zimmerman, eds. *Morality, reason and truth*. Totowa, NJ: Rowan and Allanheld, pp. 27–48.

1986. "Moral explanation of natural facts: Can moral claims be tested?" In N. Gillespie, ed. *Spindel conference 1986: Moral realism*. *Southern Journal of Philosophy* 24, Supplement: 57–68.

1987. *Change in view*. Cambridge, MA: Massachusetts Institute of Technology Press.

Hay, D. F. 1979. "Cooperative interactions and sharing between very young children and their parents." *Developmental Psychology* 15: 647–53.

Hempel, C. G. 1965. *Aspects of scientific explanation and other essays in philosophy of science*. New York: Free Press.

Hempel, C. G., and P. Oppenheim. 1948. "Studies in the logic of explanation." *Philosophy of Science* 15: 135–75.

Higgins, A., C. Power, and L. Kohlberg. 1984. "The relationships of moral atmosphere to judgments of responsibility." In W. Kurtines and J. Gewirtz, eds. *Morality, moral behavior, and moral development*. New York: Wiley, pp. 74–106.

Hoffman, M. L. 1970. "Moral development." In P. Mussen, ed. *Carmichael's manual of child psychology*, 3rd ed. New York: Wiley, pp. 261–359.

1975. "Moral internalization, parental power and the nature of the parent–child interaction." *Developmental Psychology* 11: 228–39.

1977. "Moral internalization: Current theory and research." In L. Berkowitz, ed. *Advances in experimental social psychology*. New York: Academic Press, pp. 86–135.

1981. "Is altruism part of human nature?" *Journal of Personality and Social Psychology* 40: 121–37.

1982. "Development of prosocial motivation: Empathy and guilt." In N. Eisenberg, ed. *The development of prosocial behavior*. New York: Academic Press, 281–313.

1983. "Affective and cognitive processes in moral internalization." In E. Troy, D. Higgins, N. Ruble, and S. W. Hartup, eds. *Social cognition and social development: A sociocultural perspective*. Cambridge: Cambridge University Press, pp. 236–74.

1984a. "Interaction of affect and cognition in empathy." In C. Izard, J. Kagan, and R. Zajonc, eds. *Emotion, cognition, and behavior*. New York: Cambridge University Press, pp. 103–31.

1984b. "Empathy, its limitations, and its role in a comprehensive moral theory." In W. Kurtines and J. Gewirtz, eds. *Morality, moral behavior, and moral development*. New York: Wiley, 283–302.

1988. "Moral development." In M. Bornstein and M. Lamb, eds. *Developmental psychology: An advanced textbook*. Hillsdale, NJ: Erlbaum, pp. 497–548.

References

Horgan, T., ed. 1984. "Supervenience." Spindel Conference 1983. *The Southern Journal of Philosophy* XXII, Supplement.

Hume, D. 1946. *An inquiry concerning the principles of morals.* La Salle, IL: Open Court Press. Reprinted from the 1777 edition.

1957. *An inquiry concerning human understanding.* New York: Liberal Arts Press.

Izard, C. E., and P. M. Saxton. 1988. In R. C. Atkinson, ed. *Perception and Maturation. Steven's handbook of experimental psychology,* Vol. 1, 2nd ed. New York: Wiley.

Kanfer, F., and P. Karoly. 1982. "The psychology of self-management: Abiding issues and tentative directions." In F. Kanfer and P. Karoly, eds. *Self-management and behavior change: From theory to practice.* New York: Pergamon Books, pp. 571–99.

Karoly, P. 1982. "Perspectives in self-management and behavior change." In F. Kanfer and P. Karoly, eds. *Self-management and behavior change: From theory to practice.* New York: Pergamon Books, pp. 3–31.

Kazdin, A. E. 1978. *History of behavior modification: Experimental foundations of contemporary research.* Baltimore: University Park Press.

Kettewell, H. 1973. *The evolution of melanism.* Oxford: Oxford University Press.

Kiecolt-Glaser, J. D., and R. Glaser. 1987. "Psychosocial moderators of immune function." *Annals of Behavioral Medicine* 9: 16–20.

Kim, J. 1984a. "Concepts of supervenience." *Philosophy and Phenomenological Research* 45: 153–77.

1984b. "Epiphenomenal and supervenient causation." In P. French, T. J. Uehling, and H. Wettstien, eds. *Midwest studies in philosophy,* Vol. 9. Morris: University of Minnesota Press, pp. 257–70.

1987. "'Strong' and 'global' supervenience revisited." *Philosophy and Phenomenological Research* 48: 315–26.

1989. "The myth of nonreductive materialism." *Proceedings and Addresses of the American Philosophical Association* 63: 31–47.

1990a. "Explanatory exclusion and the problem of mental causation." In E. Villaneuva, ed. *Information, semantics, and epistemology.* Oxford: Basil Blackwell, pp. 36–56.

1990b. "Supervenience as a philosophical concept." *Metaphilosophy* 21: 1–27.

1992. "'Downward causation' in emergentism and nonreductive physicalism." In A. Beckermann, H. Flohr, and J. Kim, eds. *Emergence or reduction? – Essays on the prospects of nonreductive physicalism.* New York: De Gruyter Verlag, pp. 119–38.

1993. "The nonreductivists' troubles with mental causation." In J. Heil and A. Mele, eds. *Mental causation.* Oxford: Clarendon Press, pp. 189–210.

Kincaid, H. 1986. "Reduction, explanation, and individualism." *Philosophy of Science* 53: 492–513.

1987. "Supervenience doesn't entail reducibility." *The Southern Journal of Philosophy* 25: 343–56.

1988. "Supervenience and explanation." *Synthese* 77: 251–81.

References

1990. "Molecular biology and the unity of science." *Philosophy of Science* 57: 575–93.

Kitcher, P. 1984. "1953, and all that: A tale of two sciences." *Philosophical Review* 43: 335–73.

1985. *Vaulting ambition: Sociobiology and the quest for human nature.* Cambridge, MA: Massachusetts Institute of Technology Press.

1986. "The transformation of human sociobiology." In A. Fine and P. Machamer, eds. *Proceedings of the 1986 biennial meeting of the Philosophy of Science Association.* East Lansing, MI: Philosophy of Science Association, pp. 63–74.

1987a. "Imitating selection." In S. Fox and M. Ho, eds. *Metaphors and models in evolutionary theory.* Chichester: U.K.: Wiley, pp. 295–318.

1987b. "Precis of vaulting ambition, with peer review and reply to commentators." *Behavioral and Brain Sciences* 10: 61–99.

1990. "Developmental decomposition and the future of human behavioral ecology." *Philosophy of Science* 57: 96–118.

1993. "The evolution of human altruism." *The Journal of Philosophy* XC: 497–516.

Kitcher, P., and W. C. Salmon, eds. 1989. *Scientific exploration: Minnesota studies in the philosophy of science* 13. Minneapolis: University of Minnesota Press.

Kitcher, P. S. 1984. "In defense of intentional psychology." *Journal of Philosophy* 81: 89–106.

Klee, R. L. 1984. "Micro-determinism and concepts of emergence." *Philosophy of Science* 51: 44–63.

Kohlberg, L. 1976. "Moral stages and moralization: The cognitive-developmental approach." In T. Lickona, ed. *Moral development and behavior: Theory, research and social issues.* New York: Holt, Rinehart & Winston, pp. 31–53.

1981. *The philosophy of moral development. Essays in moral development,* vol. I. San Francisco: Harper & Row.

1984. *The psychology of moral development: The nature and validity of moral stages. Essays in moral development,* Vol. II. San Francisco: Harper & Row.

Kohlberg, L., and D. Candee. 1984. "The relationship of moral judgment to moral action." In L. Kohlberg, *The psychology of moral development: The nature and validity of moral stages. Essays in moral development,* Vol. II. San Francisco: Harper & Row, pp. 426–97.

Kohlberg, L., C. Levine, and A. Hewer. 1984. "Synopses and detailed replies to critics." In L. Kohlberg, ed. *The psychology of moral development: The nature and validity of moral stages. Essays in moral development,* Vol. II. New York: Harper & Row, pp. 498–581.

Kuhn, D. 1978. "Mechanisms of cognition and social development: One psychology or two?" *Human Development* 21: 92–118.

Kurtines, W., and E. B. Grief. 1974. "The development of moral thought: Review and evaluation of Kohlberg's approach." *Psychological Bulletin* 81: 453–70.

Kurtines, W. M., and J. L. Gewirtz, eds. 1988. *Moral behavior and development: Advances in theory, research, and applications,* Vol. 1. Hillsdale, NJ: Erlbaum.

Lacey, H. 1979. "Skinner on the prediction and control of behavior." *Theory and Decision* 10: 353–85.

References

Lazarus, R. S. 1982. "Thoughts on the relations between emotion and cognition." *American Psychologist* 37: 1019–24.

1984. "On the primacy of cognition." *American Psychologist* 39: 124–9.

Leplin, J., ed. 1984. *Scientific realism.* Berkeley: University of California Press.

Lewontin, R. C. 1974. *The genetic bases of evolutionary change.* New York: Columbia University Press.

Lewontin, R. C., S. Rose, and L. Kamin. 1984. *Not in our genes.* New York: Pantheon Press.

Lickona, T. 1976. "Critical issues in the study of moral development and moral behavior." In T. Lickona, ed. *Moral development and behavior: Theory, research, and social issues.* New York: Holt, Rinehart & Winston, pp. 3–28.

Lloyd, E. 1988. *The structure and confirmation of evolutionary theory.* New York: Greenwood Press.

Lumsden, C. 1989. "Does culture need genes?" *Ethology and Sociobiology* 1: 11–28.

Lumsden, C., and E. O. Wilson. 1981. *Genes, mind and culture.* Cambridge, MA: Harvard University Press.

1983. *Promethean fire.* Cambridge, MA: Harvard University Press.

Mackie, J. L. 1977. *Ethics: Inventing right and wrong.* New York: Penguin Books.

Mahler, M. S., F. Pine, and A. Bergman. 1975. *The psychological birth of the human infant.* New York: Basic Books.

Mahoney, M. J. 1988. "The cognitive sciences and psychotherapy: Patterns in a developing relationship." In K. S. Dobson, ed. *Handbook of cognitive behavior therapies.* New York: Guilford Press, pp. 357–86.

Maier, S. F., and M. E. P. Seligman. 1976. "Learned helplessness: Theory and evidence." *Journal of Experimental Psychology: General* 105: 3–46.

Martin, G. B., and R. D. Clark. 1982. "Distress crying in neonates: Species and peer specificity." *Developmental Psychology* 18: 3–9.

Maynard-Smith, J. 1982. *Evolution and the theory of games.* Cambridge: Cambridge University Press.

Mayr, E. 1974. "Teleological and teleonomic, a new analysis." In R. S. Cohen and M. W. Wartofsky, eds. *Boston studies in the philosophy of science.* 14. Dordrecht, the Netherlands: Reidel, pp. 91–117.

McCloskey, M. 1983. "Intuitive physics." *Scientific American* 248(4): 122–30.

Meehl, P. 1950. "On the circularity of the law of effect." *Psychological Bulletin* 47: 52–75.

Menzies, P. 1988. "Against causal reductionism." *Mind* 97: 560–74.

Mischel, W., and H. N. Mischel. 1976. "A cognitive social-learning approach to morality and self regulation." In T. Lickona, ed. *Moral development and behavior: Theory, research and social issues.* New York: Holt, Rinehart & Winston, pp. 84–107.

1977. "Self-control and the self." In T. Mischel, ed. *The self: Psychological and philosophical issues.* Oxford: Basil Blackwell, pp. 31–64.

Moore, G. E. 1903. *Principia ethica.* Cambridge: Cambridge University Press.

Morgan, C. L. 1923. *Emergent evolution.* London: Willars & Morgate.

References

Murphy, J. 1982. *Evolution, morality and the meaning of life*. Totowa, NJ: Rowan and Littlefield.

Nagel, E. 1961. *The structure of science: Problems in the logic of scientific explanation*. New York: Harcourt, Brace and World.

Neurath, O. 1932. "Protokollsätze." *Erkenntnis* 3: 204–14.

Nisan, M. 1984. "Content and structure in moral judgement: An integrationist view." In W. M. Kurtines and J. L. Gewirtz, eds. *Morality, moral behavior and moral development*. New York: Wiley, pp. 208–26.

Nisbett, R., and L. Ross. 1980. *Human inference: Strategies and shortcomings of social judgment*. Englewood Cliffs, NJ: Prentice Hall.

Peirce, C. S. 1955. "How to make our ideas clear." In J. Buchler, ed. *Philosophical writings of Peirce*. New York: Dover, pp. 23–41.

Piaget, J. 1951. *Play, dreams and imitation in childhood*. New York: Norton.

Plutchik, R. 1987. "Evolutionary bases of empathy." In N. Eisenberg and J. Strayer, eds. *Empathy and its development*. New York: Cambridge University Press, pp. 38–46.

Premack, D. 1959. "Toward empirical behavior laws: I–positive reinforcement." *Psychological review* 66: 219–33.

1965. "Reinforcement theory." In D. Levine, ed. *Nebraska symposium on motivation*. Lincoln: University of Nebraska Press, pp. 123–88.

1971. "Catching up with common sense, or Two sides of a generalization." In R. Glaser, ed. *The nature of reinforcement*. New York: Academic Press, pp. 121–50.

Radke-Yarrow, M., C. Zahn-Waxler, and M. Chapman. 1983. "Children's prosocial dispositions and behavior." In M. E. Hetherington, vol. ed. *Socialization, personality and social development*. Paul H. Mussen, ed. *Handbook of child psychology*, Vol. 4, 4th ed. New York: Wiley, pp. 469–546.

Railton, P. 1986. "Moral realism." *Philosophical Review* 95: 163–207.

Rest, J. R. 1983. "Morality." In J. Flavell and E. Markmen, vol. eds. *Cognitive development*. Paul H. Mussen, eds. *Handbook of child development*, Vol. 3, 4th ed. New York: Wiley, pp. 556–629.

1984. "The major component of morality." In W. Kurtines and J. Gewirtz, eds. *Morality, moral behavior and moral development*. New York: Wiley, pp. 24–40.

Revesky, S., and J. Garcia. 1970. "Learned associations over long delays." In G. H. Bower, ed. *The psychology of learning and motivation*, Vol. 4. New York: Academic Press, pp. 1–85.

Rheingold, H. L. "Helping by two year old children." Presented at the meeting of the Society for Research in Child Development, San Francisco, March 1979.

Rheingold, H. L., D. F. Hay, and M. J. West. 1976. "Sharing in the second year of life." *Child Development* 47: 1148–58.

Richards, R. 1987. *Darwin and the emergence of evolutionary theories of mind and behavior*. Chicago: University of Chicago Press.

Ringen, J. 1976. "Explanation, teleology and operant behaviourism: A study of the experimental analysis of purposive behavior." *Philosophy of Science* 43: 223–54.

References

1985. "Operant conditioning and a paradox of teleology." *Philosophy of Science* 52: 565–77.

Rosenberg, A. 1985a. *The structure of biological science*. Cambridge: Cambridge University Press.

1985b. "Adaptationist interpretations and Panglossian paradigms." In J. H. Fetzer, ed. *Sociobiology and epistemology*. Dordrecht, the Netherlands: D. Reidel, pp. 161–80.

1986a. "Intentional psychology and evolutionary biology part I: The uncanny analogy. *Behaviorism* 14: 15–28.

1986b. "Intentional psychology and evolutionary biology part II: The crucial disanalogy." *Behaviorism* 14: 125–34.

1991. "The biological justification of ethics: A best-case scenario." *Social philosophy and policy* 8: 86–101.

Ross, H. S., and B. D. Goldman. 1977. "Attachment behavior." In T. Alloway, L. Kramer, and P. Pliner, eds. *Advances in the study of communication and affect,* Vol. 3. New York: Plenum Books, pp. 62–80.

Rottschaefer, W. A. 1976. "Observation: Theory-laden, theory-neutral, or theory free?" *Southern Journal of Philosophy* 14: 499–509.

1978. "Ordinary knowledge and scientific realism." In J. C. Pitt, ed. *The philosophy of Willfred Sellars: Queries and extensions*. Dordrecht, the Netherlands, Reidel, pp. 135–61.

1982a. "The psychological foundations of value theory: B. F. Skinner's science of values." *Zygon* 17: 293–301.

1982b. "Verbal behaviorism: An assessment of the Marras–Sellars dialogue." *Philosophy Research Archives* 9: 511–34.

1983a. "Operant learning and the scientific and philosophical foundations of behavior theory." *Behaviorism* 11: 155–62.

1983b. "Sociobiology another blow to freedom?" *Sweet Reason* 2: 63–9.

1985. "Sociobiology, religion and values: The case of E. O. Wilson." *Explorations* 4: 39–57.

1987. "Wilfrid Sellars on the nature of thought." In A. Shimony and D. Nails, eds. *Naturalistic epistemology*. Dordrecht, the Netherlands: D. Reidel, pp. 145–61.

1991a. "Evolutionary naturalistic justifications of morality: A matter of truth and works." *Biology and Philosophy* 6: 341–9.

1991b. "Social learning theories of moral agency." *Behavior and Philosophy* 19: 61–76.

1995. "B. F. Skinner and the Grand Inquisitor." *Zygon* 30: 407–34.

Rottschaefer, W. A., and D. Martinsen. 1984. "Singer, sociobiology and values: Pure reason versus empirical reason." *Zygon* 17: 159–70.

1990. "Really taking Darwin seriously: An alternative to Michael Ruse's Darwinian metaethics." *Biology and Philosophy* 5: 149–74.

Ruse, M. 1986. *Taking Darwin seriously: A naturalistic approach to philosophy*. New York: Basil Blackwell.

Ruse, M., and E. O. Wilson. 1985. "The evolution of morality." *New Scientist* 1478: 108–28.

References

1986. "Moral philosophy as applied science." *Philosophy* 61: 173–92.

Sagi, A., and M. L. Hoffman. 1976. "Empathic distress in humans." *Developmental Psychology* 12: 175–6.

Schaffner, K. F. 1993. *Discovery and explanation in biology and medicine.* Chicago: University of Chicago Press.

Schwartz, B., and H. Lacey. 1982. *Behaviorism, science and human nature.* New York: W. W. Norton.

Seligman, M. E. P. 1975. *Helplessness.* San Francisco: W. H. Freeman.

Seligman, M. E. P., and S. F. Maier. 1967. "Failure to escape traumatic shock." *Journal of Experimental Psychology* 74: 1–9.

Sellars, W. 1963. *Science, perception and reality.* New York: Humanities Press.

1968. *Science and metaphysics: Variations on Kantian themes.* New York: Humanities Press.

1979. *Naturalism and ontology.* Reseda, CA: Ridgeview.

1980. Behaviorism, language and meaning. *Pacific Philosophical Quarterly* 61: 3–25.

Simner, M. L. 1971. "Newborn responses to the cry of another infant." *Developmental Psychology* 5: 136–50.

Singer, P. 1981. *The expanding circle.* New York: Farrar, Straus and Giroux.

Skinner, B. F. 1948. *Walden two.* New York: Macmillan.

1953. *Science and human behavior.* New York: Free Press.

1971. *Beyond freedom and dignity.* New York: Knopf.

1976. *About behaviorism.* New York: Knopf.

Sober, E. 1984a. *The nature of selection: Evolutionary theory in philosophical focus.* Cambridge, MA: Massachusetts Institute of Technology Press.

ed. 1984b. *Conceptual issues in evolutionary biology.* Cambridge, MA: Massachusetts Institute of Technology Press.

1988. "What is evolutionary altruism?" In B. Linsky and M. Matthen, eds. *New essays on philosophy and biology: Canadian Journal of Philosophy supplement.* Calgary, Canada: University of Calgary Press, pp. 75–99.

1989a. "Evolutionary altruism and psychological egoism." In J. E. Fenstad, ed. *Logic, methodology and philosophy of science VIII.* New York: Elsevier Science, pp. 495–514.

1989b. "What is psychological egoism?" *Behaviorism* 17: 89–102.

1992. "The evolution of altruism – Correlation, cost, and benefit." *Biology and Philosophy* 7: 177–88.

1993. "Evolutionary altruism, psychological egoism, and morality." In M. Niteki, ed. *Evolutionary ethics.* Albany: State University of New York Press, pp. 199–216.

1994a. "Did evolution make us psychological egoists?" In E. Sober, *From a biological point of view: Essays in evolutionary philosophy.* Cambridge: Cambridge University Press, pp. 8–27.

1994b. "Prospects for an evolutionary ethics." In E. Sober, *From a biological point of view: Essays in evolutionary philosophy.* Cambridge: Cambridge University Press, pp. 93–113.

References

Sober, E., and D. S. Wilson. 1994. "A critical review of philosophical work on the units of selection problem." *Philosophy of Science* 61: 534–55.

Stump, E. 1988. "Sanctification, hardening of the heart and Frankfurt's concept of free will." *Journal of Philosophy* 8: 395–420.

Sturgeon, N. L. 1985. "Moral explanations." In D. Copp and D. Zimmerman, eds. *Morality, truth and reason.* Totowa, NJ: Rowan and Allanheld, 49–78.

1986. "Harman on moral explanations of natural facts." In N. Gillespie, ed. Spindel Conference 1986: Moral realism. *Southern Journal of Philosophy* 24: 69–78.

Suppe, F. 1989. *The semantic conception of theories and scientific realism.* Urbana: University of Illinois Press.

Taylor, C. 1977. "What is human agency?" In T. Mischel, ed. *The self: Psychological and philosophical issues.* Oxford: Basil Blackwell, pp. 103–05.

Thompson, P. 1985. "Sociobiological explanation and the testability of sociobiological theory." In J. H. Fetzer, ed. *Sociobiology and epistemology.* Dordrecht, the Netherlands: D. Reidel, pp. 201–16.

1989. *The structure of biological theories.* Albany: State University of New York Press.

Thompson, R. A. 1987. "Empathy and emotional understanding: The early development of empathy." In N. Eisenberg and J. Strayer, eds. *Empathy and its development.* New York: Cambridge University Press, pp. 119–45.

Trevarthen, C. 1982. "The primary motives for cooperation understanding." In G. Butterworth and P. Light, eds. *Social cognition: Studies in the development of understanding.* Chicago: University of Chicago Press, pp. 77–109.

Turiel, E. 1983. *The development of social knowledge: Morality and convention.* Cambridge: Cambridge University Press.

De Waal, F. 1996. *Good natured: The origins of right and wrong in humans and other animals.* Cambridge, MA: Harvard University Press.

Watson, G. 1975. "Free agency." *Journal of Philosophy* 72: 205–20.

Weinreich-Haste, H., and D. Locke. 1983. *Morality in the making.* New York: Wiley.

Williams, G. C., ed. 1975. *Group selection.* Chicago: Aldine.

1966. *Adaptation and natural selection: A critique of some current evolutionary thought.* Princeton, NJ: Princeton University Press.

1988a. "Huxley's evolution and ethics in sociobiological perspective." *Zygon* 23: 383–407.

1988b. "A reply to comments on 'Huxley's evolution and ethics in sociobiological perspective.'" *Zygon* 23: 437–8.

1989. "A sociobiological expansion of evolution and ethics." In J. Paradis and G. C. Williams, eds. *Evolution and ethics: T. H. Huxley's Evolution and ethics with new essays on its Victorian and sociobiological context.* Princeton, NJ: Princeton University Press, pp. 179–214.

Wilson, D. S., and E. Sober. 1994a. "Reintroducing group selection to the human behavioral sciences." *Behavior and Brain Sciences* 17: 585–608.

1994b. "Authors' response." *Behavior and Brain Sciences* 17: 639–47.

Wilson, E. O. 1971. *The insect societies.* Cambridge, MA: Harvard University Press.

References

1975. *Sociobiology: The new synthesis.* Cambridge, MA: Harvard University Press.

1978. *On human nature.* Cambridge, MA: Harvard University Press.

1984. *Biophilia.* Cambridge, MA: Harvard University Press.

Wren, T. 1982. "Social learning theory, self-regulation, and morality." *Ethics* 92: 409–24.

ed. 1990a. *The moral domain: Essays in the ongoing discussion between philosophy and the social sciences.* Cambridge, MA: Massachusetts Institute of Technology Press.

1990b. "The possibility of convergence between moral psychology and meta-ethics." In T. Wren, ed. *The moral domain: Essays in the ongoing discussion between philosophy and the social sciences.* Cambridge, MA: Massachusetts Institute of Technology Press, pp. 15–37.

1991. *Caring about morality: Philosophical perspectives in moral psychology.* Cambridge, MA: Massachusetts Institute of Technology Press.

Wright, L. 1976. *Teleological explanation: An etiological analysis of goals and functions.* Berkeley: University of California Press.

Zahn-Waxler, C., and M. Radke-Yarrow. 1982. "The development of altruism: Alternative research strategies." In N. Eisenberg, ed. *The development of prosocial behavior.* New York: Academic Press, pp. 109–38.

Zahn-Waxler, C., M. Radke-Yarrow, and R. A. King. 1979. "Child-rearing and children's prosocial interactions toward victims of distress." *Child Development* 50: 319–30.

Index

Index

realism (*cont.*)
  moral, integrationist theses, 218, 230–1, 238, 247
  scientific, 238–41
reduction
  and naturalistic fallacy, 232
  deductive–nomological model of, 157–61
  of manifest image, 256–60
  of mental to neurophysiological, 103–4, 148–9, 182–3n17
  predicament of, 4, 149, 156, 161–2, 179–80, 259–60, 260–3
Rest, J., 151n16
Richards, R., 51n2, 214–15n4
Rosenberg, A., 49–50
Ruse, M., 75, 214–15n4

science and ethics
  hypothesis on, *see* positions on relationships of
  hypothesis on connection, 24, 26n2, 248
  hypothesis on critical connection, 22–3, 26n2, 248, 261–3
  hypothesis on explanatory connection, 22, 25, 26n2, 248
  hypothesis on informational connection, 21
  hypothesis on meaningfulness, 24, 26n2, 248, 263–70
  hypothesis on normative connection, 23–4
  positions on relationships of, 2–3, 15–16, 25–6; *see also* science and ethics, hypothesis on
scientific image, 1, 15, 256–60, 260–2; *see also* Sellars, W.
scientific naturalism, 1–2, 11–15, 16–18, *see also* science and ethics, positions on relationships of, and hypothesis on
selection
  behavioral, 32
  cognitive, 32
  cultural, 32, 67, 83
  for, 67
  group, 35–6
  natural, 32, 67–9, 38
  of neuronal connections, 80–1
  of, 67
selfishness
  biological, 37–40
  definitional, 38
  ordinary, 37–40

self-system, 131, 132–6, 145–6, 147–9, 170–1, 215n7
Sellars, W., 1, 5–6, 163, 256–60, 272n1, 272n2, 272n3
semantic test for properties, 234–6, 250n11
semantic thesis, 233–6
separatism, *see* science and ethics, positions on relationships of
Singer, P., 214–15n4
Skinner, B. F.
  on freedom, 123n3
  on justification of moral beliefs, 196–7
  on learning history, 115–16
  on radical behaviorism, 102–5
  on science and values, 4, 98, 101, 110–12
  on science and behavior, 105–10
Sober, E., 36, 39, 47–8, 51n3
social referencing, 88, 89
sociobiology, 29, 33–7
  hypotheses on human sociobiology, 64–6
  see also Wilson, E. O.
statements
  analytic, 250n10; *see also* semantic thesis
  a priori, 250n10, 250n11
  synthetic, 250n10
  synthetic a priori, 250n10
Sturgeon, N., 242–3
supervenience, 164–73, 180–1n4, 181n6
sympathy, 86–7
synoptic vision, 1, 3, 5, 255, 260–3

tendency, *see* capacity
theories
  connectionist, of mind, 183n20
theories of moral development
  behaviorist, 83, 251n16
  cognitive developmental, 83–4, 126–30, 151n16, 251n16
  cognitive social learning, 116, 126–7, 130, 138–49
  Freduian, 83, 251n16
  Hoffman's, 86–7, 94, 206–7
  social cognitive theory, 131–5, 138–42, 149n1, 150n8; *see also* theories of moral development, cognitive social learning
theories of moral learning, *see* theories of moral development
theories of social learning, *see* theories of moral development
theory
  cognitive selection, 226, 228–30
  environmental selection, 226, 227–8